Markus Zowislok

On Moduli Spaces of Semistable Sheaves on K3 Surfaces

Markus Zowislok

On Moduli Spaces of Semistable Sheaves on K3 Surfaces

In Search of new Irreducible Symplectic Manifolds

Südwestdeutscher Verlag für Hochschulschriften

Imprint
Any brand names and product names mentioned in this book are subject to trademark, brand or patent protection and are trademarks or registered trademarks of their respective holders. The use of brand names, product names, common names, trade names, product descriptions etc. even without a particular marking in this work is in no way to be construed to mean that such names may be regarded as unrestricted in respect of trademark and brand protection legislation and could thus be used by anyone.

Publisher:
Südwestdeutscher Verlag für Hochschulschriften
is a trademark of
Dodo Books Indian Ocean Ltd., member of the OmniScriptum S.R.L Publishing group
str. A.Russo 15, of. 61, Chisinau-2068, Republic of Moldova Europe
Printed at: see last page
ISBN: 978-3-8381-1908-3

Zugl. / Approved by: Mainz, Universität, Diss., 2010

Copyright © Markus Zowislok
Copyright © 2010 Dodo Books Indian Ocean Ltd., member of the OmniScriptum S.R.L Publishing group

To my father

Contents

Introduction		**iii**
1	**State of the art**	**1**
	1.1 Symplectic varieties	1
	1.2 Symplectic resolutions	2
	1.3 Sheaves on K3 surfaces	4
	1.4 General ample divisors	5
	1.4.1 Walls for two-dimensional sheaves	6
	1.4.2 Walls for one-dimensional sheaves	9
	1.5 Semistable sheaves on K3 surfaces	13
	1.6 Moduli spaces of sheaves on K3 surfaces	14
2	**Irreducible components**	**19**
	2.1 A decomposition	19
	2.2 Products and symmetric products	21
	2.3 Symplectic resolvability of components	23
3	**Moduli spaces for one-dimensional sheaves**	**27**
	3.1 Morphisms between moduli spaces	27
	3.2 Results	31
4	**Moduli spaces for (H, A)-semistable sheaves**	**33**
	4.1 Preliminaries	33
	4.2 Semistable sheaves	34

4.3		Jordan-Hölder filtration and S-equivalence	42
4.4		Flat families	43
4.5		The moduli functor	44
4.6		The construction of the moduli space	45
4.7		The construction - Proofs	49
4.8		Local properties and dimension estimates	54
4.9		Universal families	55

5 Moduli spaces for two-dimensional sheaves — 57

5.1	Semistable sheaves	57
5.2	The moduli space $M_{H,A}(v)$	62
5.3	Terminalisations for $M_H(v)$	67
5.4	Birational maps between moduli spaces	70
5.5	On the discriminant	71
5.6	Existence of stable sheaves	75
5.7	More results on $M_{H,A}(v)$	84
5.8	Results on $\overline{M_H^s(v)}$	86

6 The relation to twisted stability — 89

6.1	Twisted stability	89
6.2	Two-dimensional sheaves on a K3 surface	91

Bibliography — 95

Acknowledgement — 99

Introduction

A compact complex manifold of Kähler type is Ricci flat if and only if its first Chern class is trivial by [Yau78]. The structure of such manifolds is described by the Bogomolov decomposition (see [Bea83] Theorem 2, [Ber55] or [Bog74]):

Theorem. *Let X be a compact manifold of Kähler type with trivial first Chern class. Then there is a finite étale covering X' of X, which is isomorphic to the product*
$$T \times \prod V_i \times \prod X_j,$$
where each V_i is a projective simply connected manifold of dimension $m_i \geq 3$ with trivial canonical sheaf and $H^0(V_i, \Omega_{V_i}^p) = 0$ for $0 < p < m_i$, each X_j is a compact irreducible symplectic manifold of Kähler type and T is a complex torus.

The only compact irreducible simply connected Kähler manifolds with trivial first Chern class hence are compact irreducible symplectic manifolds and Calabi-Yau manifolds.

There is a recent publication [Bea10] on holomorphic symplectic geometry we want to advert to.

The compact irreducible symplectic manifolds are exactly the irreducible compact hyperkähler manifolds ([Huy99] or [Bea83]). The symplectic form can easily be defined on an irreducible compact hyperkähler manifold but conversely there is only a pure existence result for a hyperkähler metric on a compact irreducible symplectic manifold based on Yau's solution of the Calabi conjecture.

Up to now only very few examples of compact irreducible symplectic manifolds are known, these are up to deformation:

- K3 surfaces

- Hilbert schemes of points on a K3 surface

- generalised Kummer varieties associated to an abelian surface

- a 10-dimensional example of O'Grady [O'G99]

- a 6-dimensional example of O'Grady [O'G03]

All examples above can be constructed from moduli spaces of semistable sheaves on K3 or abelian surfaces. E.g. the Hilbert scheme $\text{Hilb}^n(X)$ over a K3 surface X is isomorphic to the moduli space of sheaves on X with rank 1, $c_1 = 0$ and $c_2 = n$. By [KLS06] it is almost completely excluded that this construction yields other examples than those mentioned above - proposition 1.6.4 summarises known results.

We want to investigate the missing cases but restrict our considerations to sheaves on K3 surfaces. What is left here are the moduli spaces of semistable sheaves with rank 0, given first Chern class, and Euler characteristic 0, and the moduli spaces $M_H(v)$ of sheaves with Mukai vector v that are semistable with respect to a not v-general ample divisor H.

In chapter 1 we recall the state of the art. We give the definition of a symplectic variety and cite the important result of Huybrechts that birational projective irreducible symplectic manifolds are deformation equivalent. We recall the definition of the Mukai vector of a sheaf on a K3 surface together with first properties and discuss the notion of a general ample divisor. The last section collects known facts on moduli spaces of semistable sheaves on K3 surfaces and their symplectic resolvability.

In chapter 2 we investigate irreducible components of the moduli space $M_H(v)$. Our first result gives a relation to components of other moduli spaces containing stable sheaves:

Proposition. (2.1.1) *Let X be a projective K3 surface, H an ample divisor, $v \in \Lambda(X) := \mathbb{N}_0 \oplus \mathrm{NS}(X) \oplus \mathbb{Z} \subset H^{2*}(X, \mathbb{Z})$ and M an irreducible component of $M_H(v)$. Then there is a birational projective morphism*

$$g : \prod_{i=1}^{m} S^{n_i} M_i \to M$$

for a suitable decomposition $v = \sum_{i=1}^{m} n_i v_i$ with $n_i \in \mathbb{N}$ and $v_i \in \Lambda(X)$ for $1 \leq i \leq m$ and a suitable choice of pairwise distinct irreducible components $M_i \subset \overline{M_H^s(v_i)}$ for $1 \leq i \leq m$.

This leads us to a section on products and symmetric products of symplectic varieties. Furthermore, we generalise an important result of Namikawa:

Proposition. (2.3.4) *If there is a singular $\mathbb{Q}$-factorial projective symplectic terminalisation of a projective scheme X then X admits no projective symplectic resolution.*

Therefore we will always try to find a $\mathbb{Q}$-factorial projective symplectic terminalisation. The chapter ends with the following result:

Theorem. (2.3.5) *Assume the situation of proposition 2.1.1.*

1. *Assume that for each $1 \leq i \leq m$ there is a $\mathbb{Q}$-factorial projective symplectic terminalisation $\tilde{M}_i \to M_i$.*

   a) *Let $\tilde{M}_i$ be nonsingular for all $1 \leq i \leq m$ and let $v_i^2 \leq 0$ for all $1 \leq i \leq m$*

with $n_i > 1$. Then there is a projective symplectic resolution $\tilde{M} \to M$. If $\tilde{M}$ can be chosen to be an irreducible symplectic manifold then it is deformation equivalent to $\tilde{M}_i$ for some $1 \leq i \leq m$ or to a Hilbert scheme of points on a K3 surface.

b) Let $\tilde{M}_j$ be singular or let $v_j^2 \geq 2$ and $n_j > 1$ for some j with $1 \leq j \leq m$. Then there is a singular $\mathbb{Q}$-factorial projective symplectic terminalisation $\tilde{M} \to M$. In particular, M admits no projective symplectic resolution.

2. Let $U := g(\prod_{i=1}^m S^{n_i} M_i^s)$ and U' be the normalisation of U_{red}. Then there is a $\mathbb{Q}$-factorial symplectic terminalisation $\tilde{U} \to U$ and U' is a $\mathbb{Q}$-factorial symplectic variety.

Moreover, if $v_j^2 \geq 2$ and $n_j > 1$ for some j with $1 \leq j \leq m$ then there is no projective symplectic resolution of M.

Chapter 3 gives a complete answer to our question for moduli spaces of one-dimensional semistable sheaves on K3 surfaces. We follow the idea of constructing a $\mathbb{Q}$-factorial symplectic terminalisation. First we discuss how one can construct morphisms between moduli spaces when the ample divisor varies. The Mukai vector $(0, c, 0)$ is usually excluded from considerations because there is no notion of a general ample divisor. We are able to reduce these special cases of $M_H(0, c, 0)$ to those we are able to treat later:

Theorem. (3.1.4) *Let X be a K3 surface with an ample divisor H and $0 \neq c \in H^2(X, \mathbb{Z})$ effective. Then there is an isomorphism*

$$M_H(0, c, 0) \cong M_H(0, c, c.H),$$

which is induced by tensoring with H, and one has $c.H > 0$.

So let $v = (0, v_1, v_2)$ with $v_1 \neq 0$ effective and $v_2 \neq 0$. Then one can construct a morphism $M_A(v) \to M_H(v)$ for nongeneral H choosing some general A near

H. $M_A(v)$ is known to be symplectically resolvable or $\mathbb{Q}$-factorial symplectic with terminal singularities. Thus we are able to answer our main question for $\overline{M_H^s(v)}$ and therefore for any component of $M_H(v)$:

Corollary. (3.2.3) *Let X be a projective K3 surface, $v = (0, v_1, v_2) \in \Lambda(X)$ with $v_1 \neq 0$ effective, H an ample divisor on X and M an irreducible component of $M_H(v)$. If there is a projective symplectic resolution $\tilde{M} \to M$ with $\tilde{M}$ an irreducible symplectic manifold then it is deformation equivalent to a symplectic resolution of some $M_A(w)$, where $w \in \Lambda(X)$ and A is some w-general ample divisor.*

We would like to work out the same steps for the case of moduli spaces of two-dimensional semistable sheaves on K3 surfaces. Unfortunately, in the case of sheaves of positive rank an H-semistable sheaf for general H might become H-unstable when H is moved onto a so-called wall, i.e. when H becomes nongeneral, and there is no morphism between the corresponding moduli spaces in general. We need to find another way. In [MW97] the authors construct a moduli space for twisted semistable sheaves of fixed Chern character on a surface. They show under certain conditions an equivalence of twisted semistability and semistability with an extra condition involving a second ample divisor A. We give the latter one the name (H, A)-semistability. Although in [MW97] the moduli space for (H, A)-semistable sheaves is constructed, we need much more details and properties, so we give another construction following the one in the book [HL97] in order to generalise results therein and others. This is the topic of chapter 4. In our application our semistable sheaves will be living on a K3 surface but as there seems to be no reason for such a restriction we stay as general as possible without getting into unnecessary trouble. This chapter is heavily based on the book [HL97]. After some preparatory facts we introduce the notion of (H, A)-(semi)stability, where H and A are two ample line bundles on a projective scheme. As the new notion of (H, A)-(semi)stability includes the notion of H-(semi)stability, we automatically recall the latter one and some of its properties. The definition immediately yields the observation

$$H\text{-stable} \Rightarrow (H,A)\text{-stable} \Rightarrow (H,A)\text{-semistable} \Rightarrow H\text{-semistable}.$$

Therefore we can get the needed morphisms between the corresponding moduli spaces. We give different characterisations of (H, A)-(semi)stability, explain the Jordan-Hölder filtration and prove the openness property of (H, A)-(semi)stability in flat families. Then the construction follows. The generalisation in the construction is made by using two different ample line bundles H and A, the first one in order to make the considered sheaves globally generated, and the second one in order to get the linearised line bundle. We prove:

Theorem. (4.6.5) *There is a projective coarse moduli space $M_{H,A}(P)$ for (H, A)-semistable sheaves with fixed Hilbert polynomial P with respect to H.*

Almost all proofs carry over literally replacing the expression H-(semi)stable by the expression (H, A)-(semi)stable. The most interesting part might be the equivalence of (semi)stable points (in the GIT sense) and (semi)stable sheaves. This is the place where one sees where the definition of (H, A)-semistability comes from. Finally we discuss some local properties and deduce the existence of a quasi-universal family on the stable locus.

In chapter 5 we leave the generality and return to our original question for moduli spaces of torsion free semistable sheaves. We first calculate explicit expressions for the semistability condition on a nonsingular projective surface. Then we further restrict to K3 surfaces and show that by choosing A general one can force destabilising subsheaves of an (H, A)-semistable sheaf to have a Mukai vector proportional to the Mukai vector of the given sheaf, which is the key fact of the analysis of moduli spaces of sheaves on K3 surfaces for general ample divisors in [KLS06]. The moduli space $M_{H,A}(v)$ of (H, A)-semistable sheaves with Mukai vector v is constructed analogously to the case of H-semistable sheaves by taking the fibre of the determinant morphism. We explain which properties of $M_H(v)$ carry over immediately to $M_{H,A}(v)$, this time in particular with v-general A, and get the following result:

Theorem. (5.2.5) Let X be a projective K3 surface, $v = (v_0, v_1, v_2) \in \Lambda(X)$ primitive with $v_0 > 0$, $m \in \mathbb{N}$, H an ample divisor on X and A an mv-general ample divisor on X. Assume that $M^s_{H,A}(mv)$ is nonempty. Then $v^2 \geq -2$.

1. If $v^2 = -2$ then $m = 1$ and $M_{H,A}(v)$ consists of a reduced point.

2. If $v^2 = 0$ then $M_{H,A}(mv) = M^s_{H,A}(mv)$, and $M_{H,A}(mv)$ is a projective symplectic nonsingular surface.

3. Let $v^2 \geq 2$ and $M^s_{H,A}(v)$ be nonempty. Then $M_{H,A}(mv)$ is a projective symplectic variety of dimension $2 + m^2 v^2$.

    a) If $m = 1$ then $M_{H,A}(v) = M^s_{H,A}(v)$, and $M_{H,A}(v)$ is nonsingular.

    b) If $m \geq 2$ then the singular locus of $M_{H,A}(mv)$ is nonempty and equals the strictly semistable locus.

        i. If $m = 2$ and $v^2 = 2$ then the singular locus has codimension 2 and $M_{H,A}(mv)$ admits a symplectic resolution.

        ii. If $m = 2$ and $v^2 > 2$ or $m > 2$ then $M_{H,A}(mv)$ is locally factorial, the singular locus has codimension at least 4 and the singularities are terminal. There is no open neighbourhood of a singular point that admits a symplectic resolution.

4. Let $v^2 \geq 2$ but now $M^s_{H,A}(v)$ be empty. Then $M_{H,A}(v)$ is empty as well, i.e. $m > 1$ by assumption. If $m = 2$ or 3 then $M_{H,A}(mv) = M^s_{H,A}(mv)$, and $M_{H,A}(mv)$ is a nonsingular projective symplectic variety of dimension $2 + m^2 v^2$.

The assumption that $M^s_{H,A}(mv)$ is nonempty is not crucial: in our application this holds automatically. But the assumption that $M^s_{H,A}(v)$ is nonempty for $v^2 \geq 2$ and $m > 2$ or for $v^2 > 2$ and $m \geq 2$ is problematic. Not only that the lacking of such sheaves kills our terminalisation, this might also produce unexplored possibly nonsingular or symplectically resolvable symplectic varieties, in particular as stated in the last item. Another question is whether the nonsingular projective symplectic varieties are deformation equivalent to known examples - of course, the surface case is not interesting as surfaces are classified. Because of theorem 1.1.4 it is enough to establish birational equivalence. Anyway the moduli spaces $M_{H,A}(v)$ and their symplectic resolutions are good candidates for $\mathbb{Q}$-factorial projective symplectic terminalisations of $M_H(v)$, and we can reduce our question on $M_H(v)$ to the investigation of $M_{H,A}(v)$:

Proposition. (5.3.2) *Let X be a projective K3 surface, $v = (v_0, v_1, v_2) \in \Lambda(X)$ with $v_0 \geq 2$, H a not v-general and A a v-general ample divisor. Assume that $M_H^s(v)$ is nonempty and that there is a $\mathbb{Q}$-factorial projective symplectic terminalisation $\tilde{M}_{H,A}(v) \to M_{H,A}(v)$. Then there is a $\mathbb{Q}$-factorial projective symplectic terminalisation $f : \tilde{M}_{H,A}(v) \to \overline{M_H^s(v)}$.*

1. *If there is a projective symplectic resolution $\tilde{M} \to \overline{M_H^s(v)}$ and $\tilde{M}$ can be chosen to be an irreducible symplectic manifold then $\tilde{M}$ is unique up to deformation.*

2. *If $\tilde{M}_{H,A}(v)$ is singular then $\overline{M_H^s(v)}$ admits no projective symplectic resolution.*

Corollary. (5.3.3) *Let X be a projective K3 surface, $v = (v_0, v_1, v_2) \in \Lambda(X)$ with $v_0 \geq 2$, H a not v-general ample divisor on X and M an irreducible component of $M_H(v)$ containing no H-stable sheaves.*

Assume that for all $w = (w_0, w_1, w_2) \in \Lambda(X)$ with $1 < w_0 < v_0$ and such that H is not w-general, $\frac{w_1.H}{w_0} = \frac{v_1.H}{v_0}$ and $\frac{w_2}{w_0} = \frac{v_2}{v_0}$ there is a $\mathbb{Q}$-factorial projective symplectic terminalisation of $M_{H,A_w}(w)$ for a suitable w-general ample divisor A_w. Then there is a $\mathbb{Q}$-factorial projective symplectic terminalisation $\tilde{M} \to M$.

If $\tilde{M}$ can be chosen to be an irreducible symplectic manifold then it is deformation equivalent to some symplectic resolution of some $M_{H,A}(w)$, where $w = (w_0, w_1, w_2) \in \Lambda(X)$ has the above properties and A is a w-general ample divisor, to a symplectic resolution of some $M_H(w)$, where $1 \leq w_0 < v_0$, H is w-general, $\frac{w_1.H}{w_0} = \frac{v_1.H}{v_0}$ and $\frac{w_2}{w_0} = \frac{v_2}{v_0}$, or to a Hilbert scheme of points on a K3 *surface.*

Unfortunately, theorem 5.2.5 does not give a complete answer and the assumptions are not necessarily satisfied. The next section reduces the gaps to the existence of certain stable sheaves. Hence we need existence results for stable sheaves. After some preparatory calculations we can state the numerical condition

$$v^2 > 2\left(v_0^3 - v_0^2 - v_0 - (v_0 - 1)\left\lfloor \frac{v_0^2}{4} \right\rfloor^{-1}\right) =: \varphi(v_0),$$

which ensures the existence of the needed stable sheaves. After a separate treatment of the case $v^2 = 2 = \varphi(2)$ with $v_0 = 2$ we get:

Theorem. (5.7.1) *Let X be a projective K3 surface, $v = (v_0, v_1, v_2) \in \Lambda(X)$ primitive and $m \in \mathbb{N}$ with $mv_0 \geq 2$, H a not mv-general ample divisor on X and A an mv-general ample divisor on X in a chamber touching H.*

1. *If $(mv)^2 > \varphi(mv_0)$ then $M_{H,A}^s(mv)$ is nonempty.*

2. *Let $m = 1$ and assume $v^2 > \varphi(v_0)$. Then $M_{H,A}(v)$ is an irreducible symplectic manifold and deformation equivalent to $\text{Hilb}^{\frac{v^2}{2}+1}(X)$.*

3. *Let $m = 1$ and $v = (2, v_1, \frac{v_1^2-2}{4})$. Then $M_{H,A}(v)$ is birational to $\text{Hilb}^2(X)$ or to X^2 or it is empty. In the first case $M_{H,A}(v)$ is an irreducible symplectic manifold and deformation equivalent to $\text{Hilb}^2(X)$, in the second case it cannot be an irreducible symplectic manifold.*

4. *Let $m = 2$, $v^2 = 2$ and $v_0 = 1$. Then the symplectic varieties $M_{H,A}(2v)$ and $M_B(2v)$ are birational for a suitable $2v$-general ample divisor B on X, hence also any symplectic resolutions M of $M_{H,A}(2v)$ and M' of $M_B(2v)$.*

   If furthermore M or M' is an irreducible symplectic manifold then both are irreducible symplectic and deformation equivalent.

5. Let $M^s_{H,A}(mv)$ be nonempty, $m \geq 2$ and $(mv)^2 \geq 16$, and assume $v_0 = 1$ or $v^2 > \varphi(v_0)$. Then $M_{H,A}(mv)$ is a singular locally factorial (and therefore $\mathbb{Q}$-factorial) projective symplectic variety with only terminal singularities, and in particular, there is no projective symplectic resolution.

In the last section we return to H-semistable sheaves on K3 surfaces and deduce the following:

Theorem. (5.8.1) *Let X be a projective K3 surface, $v = (v_0, v_1, v_2) \in \Lambda(X)$ primitive with $v_0 > 0$, $m \in \mathbb{N}$ and H an ample divisor on X. Furthermore, assume that $M^s_H(mv)$ is nonempty. Then one has $v^2 \geq -2$, and in the case of equality one has $m = 1$ and $M_H(v)$ consists of a reduced point. Let now $v^2 \geq 0$.*

1. *Let $m = 1$ or $(mv)^2 \leq 14$. Then there is a projective symplectic resolution $M \to \overline{M^s_H(mv)}$. If H is not mv-general then M can be chosen to be a symplectic resolution of $M_{H,A}(mv)$, where A is an mv-general ample divisor.*

   Moreover, if M can be chosen to be irreducible symplectic then it is unique up to deformation.

2. *Let $m \geq 2$ and $(mv)^2 \geq 16$. If H is mv-general or $v_0 = 1$ or $v^2 > \varphi(v_0)$ then there is a singular locally factorial (and therefore $\mathbb{Q}$-factorial) projective symplectic terminalisation of $\overline{M^s_H(mv)}$, and in particular, there is no projective symplectic resolution of $\overline{M^s_H(mv)}$.*

We conclude the section with a discussion for small ranks and point out that the moduli spaces of semistable rank two sheaves do not yield new examples of irreducible symplectic manifolds.

Finally, in chapter 6, we discuss the relation of (H, A)-stability and twisted stability.

Chapter 1

State of the art

In this chapter we recall the state of the art. We assume familiarity with the book [HL97]. Our ground field for this chapter is $\mathbb{C}$.

1.1 Symplectic varieties

We recall the definition of a symplectic variety following [Bea00].

Definition 1.1.1 *1. Let X be a variety. A nonsingular variety $\tilde{X}$ together with a proper birational morphism $f : \tilde{X} \to X$ is called a resolution of singularities if it induces an isomorphism $f^{-1}(U) \xrightarrow{\cong} U$, where U is the nonsingular locus of X.*

2. A closed nondegenerate holomorphic 2-form on a nonsingular variety is called a (holomorphically) symplectic form.

3. A symplectic variety X is a normal variety together with a symplectic form ω on the nonsingular locus U of X such that there is a resolution of singularities $f : \tilde{X} \to X$ for which the pullback $\left(f|_{f^{-1}(U)}\right)^* \omega$ extends to a holomorphic 2-form on $\tilde{X}$.

One can show that if X is a symplectic variety and $f : \tilde{X} \to X$ is any resolution of singularities then the pullback of ω by the induced isomorphism extends to a holomorphic 2-form on $\tilde{X}$, see [Bea00] section 1.

Definition 1.1.2 *An irreducible symplectic manifold is a simply connected compact Kähler manifold with a holomorphically symplectic form that generates* $H^0(X, \Omega_X^2)$.

Theorem 1.1.3 *Let X be a projective K3 surface. Then* $\mathrm{Hilb}^m(X)$ *is a projective irreducible symplectic manifold of dimension $2m$.*

Proof. [Bea83] theorem 3. $\square$

Theorem 1.1.4 *Let a projective irreducible symplectic manifold X and a projective manifold X' carrying a nondegenerate 2-form be birationally equivalent. Then X' is irreducible symplectic, too, and X and X' are deformation equivalent and hence diffeomorphic.*

Proof. X' is irreducible symplectic by the proof of [HL97] corollary 6.2.7, and X and X' are deformation equivalent and diffeomorphic by [Huy99]. $\square$

1.2 Symplectic resolutions

Definition 1.2.1 1. *If a morphism between schemes induces an isomorphism between open dense subsets then we say that the morphism is birational.*

2. *Let X be a scheme. A nonsingular symplectic variety $\tilde{X}$ together with a proper birational morphism $f : \tilde{X} \to X$ is called a symplectic resolution.*

Note that we do not require f to be an isomorphism over the nonsingular locus. For a projective symplectic resolution of a projective normal variety the following proposition shows that this condition always holds true. Moreover, if in this case ω is the symplectic form on the nonsingular locus of X induced by f then the pullback of ω clearly extends to the original symplectic form on $\tilde{X}$. Note that this is the usual definition for a resolution of singularities $f : \tilde{X} \to X$ of a symplectic variety X to be symplectic.

1.2. SYMPLECTIC RESOLUTIONS

Proposition 1.2.2 *Let $f : \tilde{X} \to X$ be a projective symplectic resolution of a projective scheme X. Then*

1. *X is irreducible,*

2. *the normalisation X' of X_{red} is a projective symplectic variety,*

3. *f factors through a projective symplectic resolution $h : \tilde{X} \to X'$ and*

4. *h induces an isomorphism $h^{-1}(X'_{sm}) \xrightarrow{\cong} X'_{sm}$, where X'_{sm} is the nonsingular locus of X'.*

Proof. As $\tilde{X}$ is irreducible, so is $f(\tilde{X}) = X$. Hence X' is a normal integral scheme, which is projective by [Mum99] theorem III.8.4, i.e. X' is a projective normal variety.

The morphism f factors through X' and hence yields a projective birational morphism $h : \tilde{X} \to X'$. In particular, there is an open dense subset $U \subset \tilde{X}$ such that h induces an isomorphism $U \to h(U)$. We maximally extend its inverse and get a morphism $i : V \to \tilde{X}$ with $h(U) \subset V$ and $(h \circ i)|_{h(U)} = \mathrm{id}_V|_{h(U)}$. Then by [Har77] lemma I.4.1 one already has $h \circ i = \mathrm{id}_V$. In particular, i is injective and induces an isomorphism $V_{sm} \xrightarrow{\cong} i(V_{sm})$ where V_{sm} denotes the nonsingular locus of V, and we are able to pull back the symplectic form $\omega_{\tilde{X}}$ on $\tilde{X}$ to a symplectic form $\omega_{V_{sm}}$ on V_{sm}. By [Har77] lemma V.5.1 one has $\mathrm{codim}_{X'}(X' \setminus V) \geq 2$ as X' is normal, so one has $\mathrm{codim}_{X'_{sm}}(X'_{sm} \setminus V_{sm}) \geq 2$ for the nonsingular loci, and the form $\omega_{V_{sm}}$ can be extended to a closed 2-form $\omega_{X'_{sm}}$ on X'_{sm}. The degeneration locus $D := \{p \in X'_{sm} \mid (\omega_{X'_{sm}})_p \text{ is degenerate}\}$ is a divisor and must be contained in $X'_{sm} \setminus V_{sm}$, which has codimension at least 2, and therefore D is empty, i.e. $\omega_{X'_{sm}}$ is a symplectic form on X'_{sm}.

Let $\tilde{X}^o := h^{-1}(X'_{sm})$, $h^o : \tilde{X}^o \to X'_{sm}$ be the restriction of h to $\tilde{X}^o$ and $\omega := (h^o)^* \omega_{X'_{sm}}$, which is a closed 2-form on $\tilde{X}^o$. The degeneration locus $\{p \in \tilde{X}^o \mid \omega_p \text{ is degenerate}\}$ is the canonical divisor, which is trivial as $\tilde{X}^o$ is a symplectic variety, hence ω is a symplectic form on $\tilde{X}^o$. Assume that h^o is not an isomorphism, so there is a $p \in X'_{sm}$ such that $Z := (h^o)^{-1}(\{p\})$ has dimension at least 1, and ω degenerates on Z, a contradiction. Thus h^o is an isomorphism. □

Corollary 1.2.3 *Let* $f : \tilde{X} \to X$ *be a projective symplectic resolution of a projective normal variety* X. *Then* f *is a resolution of singularities.*

1.3 Sheaves on K3 surfaces

We skip introducing K3 surfaces and refer to [BHPVdV04]. Let X be a K3 surface and E and F coherent sheaves on X. One has $\chi(\mathcal{O}_X) = 2$ and hence $\mathrm{td}(X) = (1, 0, 2) \in H^{2*}(X, \mathbb{Z})$.

Definition 1.3.1 *The Mukai vector of* E *is* $v(E) := \mathrm{ch}(E)\sqrt{\mathrm{td}(X)}$.

Mukai calls it the vector associated to E, see [Muk87] definition 2.1. By Riemann-Roch one has $\chi(E) = \mathrm{ch}_2(E) + 2\mathrm{rk}\, E$, hence

$$v(E) = (\mathrm{rk}\, E, c_1(E), \chi(E) - \mathrm{rk}\, E) \in \mathbb{N}_0 \oplus \mathrm{NS}(X) \oplus \mathbb{Z}.$$

Notation 1.3.2 $\Lambda(X) := \mathbb{N}_0 \oplus \mathrm{NS}(X) \oplus \mathbb{Z} \subset H^{2*}(X, \mathbb{Z})$.

In [Muk87] section 2, Mukai defines a symmetric integral bilinear form on $\Lambda(X)$ by

$$\langle (v_0, v_1, v_2), (v'_0, v'_1, v'_2) \rangle := v_1.v'_1 - v_0 v'_2 - v'_0 v_2,$$

which is now called the Mukai pairing.

Proposition 1.3.3

$$\chi(E, F) := \hom(E, F) - \mathrm{ext}^1(E, F) + \mathrm{ext}^2(E, F) = -\langle v(E), v(F) \rangle.$$

Proof. This follows from Riemann-Roch. □

Proposition 1.3.4 (Mukai) *The pairing*

$$\mathrm{Ext}^i(E,F) \times \mathrm{Ext}^{2-i}(F,E) \to H^2(\mathcal{O}_X), \ (a,b) \mapsto \mathrm{tr}^2(a \circ b)$$

is nondegenerate for every i. In particular, $\mathrm{ext}^1(E,F) = \mathrm{ext}^1(F,E)$ and $\mathrm{ext}^2(E,F) = \hom(F,E)$.

Proof. [Muk87] proposition 2.3. □

Corollary 1.3.5 $\langle v(E), v(F) \rangle = \mathrm{ext}^1(E,F) - \hom(E,F) - \hom(F,E)$. *In particular, $v(E)^2 := \langle v(E), v(E) \rangle = \mathrm{ext}^1(E,E) - 2\mathrm{end}(E)$. If E is simple, i.e. $\mathrm{end}(E) = 1$, then $v(E)^2 \geq -2$.*

Definition 1.3.6 *A vector v of a lattice Λ is primitive or indivisible if there is no decomposition $v = mw$ with $2 \leq m \in \mathbb{N}$ and $w \in \Lambda$.*

Proposition 1.3.7 (Mukai) *Let H be an ample divisor on X, $v \in \Lambda(X)$ primitive, $m \in \mathbb{N}$ and E an H-stable sheaf with $v(E) = mv$. Then $v^2 \geq -2$.*

Moreover, if $v^2 = -2$ then $m = 1$, and if additionally F is any H-semistable sheaf with $v(F) = v$ then $F \cong E$.

Proof. By [HL97] corollary 1.2.8 the sheaf E is simple, hence $m^2 v^2 \geq -2$ by corollary 1.3.5. Thus $v^2 \geq -\frac{2}{m^2}$, i.e. either $v^2 = -2$ and $m = 1$, or $v^2 \geq 0$.

Assume that $v^2 = -2$ and that there is an H-semistable sheaf F with $v(F) = v$. By proposition 1.3.3 one has

$$2 = -\langle v(E), v(F) \rangle = \chi(E,F) = \hom(E,F) + \hom(F,E) - \mathrm{ext}^1(E,F),$$

hence $\hom(E,F) > 0$ or $\hom(F,E) > 0$, and by [HL97] proposition 1.2.7 any such homomorphism is an isomorphism. □

1.4 General ample divisors

We recall the notion of a general ample divisor and explain its advantage.

1.4.1 Walls for two-dimensional sheaves

We follow the presentation in [HL97] section 4.C. Let X be a nonsingular projective surface over an algebraically closed field k of characteristic zero, and $\mathrm{Num}(X) := \mathrm{Pic}(X)/\equiv$, where $\equiv$ denotes numerical equivalence. Let $r \geq 2$ and $\Delta > 0$ be integers.

Definition 1.4.1 *Let*

$$W(r, \Delta) := \{\xi^\perp \cap \mathrm{Amp}(X)_\mathbb{Q} \mid \xi \in \mathrm{Num}(X) \quad \text{with} \quad -\frac{r^2}{4}\Delta \leq \xi^2 < 0\},$$

whose elements are called walls. The connected components of the complement of the union of all walls are called chambers. An ample divisor is called general if it is not contained in a wall.

If X is a K3 surface and $v = (v_0, v_1, v_2) \in \Lambda(X)$ with $\Delta = v^2 + 2v_0^2$ (this is the discriminant of a sheaf with Mukai vector v) then we also write more precisely v-general instead of general. Furthermore, in the case of $v_0 = 1$ we agree that all divisors are v-general.

Lemma 1.4.2 *The set $W(r, \Delta)$ is locally finite in $\mathrm{Amp}(X)_\mathbb{Q}$.*

Proof. [HL97] lemma 4.C.2. □

Theorem 1.4.3 *Let H be an ample divisor, F a μ_H-semistable coherent sheaf of rank r and discriminant Δ and $F' \subset F$ a subsheaf of rank r', $0 < r' < r$, with $\mu_H(F') = \mu_H(F)$. Then $\xi := r.c_1(F') - r'c_1(F)$ satisfies*

$$\xi.H = 0 \quad \text{and} \quad -\frac{r^2}{4}\Delta \leq \xi^2 \leq 0,$$

and $\xi^2 = 0$ if and only if $\xi = 0$.

Proof. [HL97] theorem 4.C.3. □

1.4. GENERAL AMPLE DIVISORS

The proof of the theorem also works for $\Delta = 0$ and then yields $\xi = 0$. This explains why there are no walls for $\Delta = 0$, so we assumed $\Delta > 0$ in the beginning of this section, the case of $\Delta < 0$ being excluded by the Bogomolov inequality (see e.g. [HL97] theorem 3.4.1).

Corollary 1.4.4 *Let the situation be as in the theorem and let $\mu_A(F') = \mu_A(F)$ for some general ample divisor A. Then $\xi = 0$.*

Proof. Assume $\xi \neq 0$. By the theorem one has $-\frac{r^2}{4}\Delta \leq \xi^2 < 0$, thus ξ defines a wall containing H and A in contradiction to A being general. □

Lemma 4.C.5 of [HL97] can be generalised:

Lemma 1.4.5 *Let H and H' be two ample divisors on X and F a torsion free sheaf on X that is μ_H-semistable but not $\mu_{H'}$-semistable. Then there is an ample $\mathbb{Q}$-divisor $H_0 \in [H, H'[= \{(1-t)H + tH' \mid t \in [0,1[\}$ and a nontrivial proper saturated subsheaf $F_0 \subset F$ such that F and F_0 are μ_{H_0}-semistable with $\mu_{H_0}(F_0) = \mu_{H_0}(F)$ and $\mu_{H'}(F_0) > \mu_{H'}(F)$.*

Proof. If there is a nontrivial proper saturated subsheaf $F_0 \subset F$ with $\mu_H(F_0) = \mu_H(F)$ and $\mu_{H'}(F_0) > \mu_{H'}(F)$ then we can choose $H_0 = H$. So we can assume that $\mu_H(F') < \mu_H(F)$ for all nontrivial proper saturated subsheaves $F' \subset F$ with $\mu_{H'}(F') > \mu_{H'}(F)$. This is the situation in the proof of [HL97] lemma 4.C.5, so it carries over literally. Note that strictly speaking the proof does not construct an $H_0 \in [H, H'[$ but rather a multiple of it, which, of course, has no effect. □

The original lemma 4.C.5 of [HL97] is now an immediate consequence:

Corollary 1.4.6 *Let H and H' be two ample divisors on X and F a torsion free sheaf on X that is μ_H-stable but not $\mu_{H'}$-stable. Then there is an ample $\mathbb{Q}$-divisor $H_0 \in]H, H']$ and a nontrivial proper saturated subsheaf $F_0 \subset F$ such that F and F_0 are μ_{H_0}-semistable with $\mu_{H_0}(F_0) = \mu_{H_0}(F)$ and $\mu_{H'}(F_0) \geq \mu_{H'}(F)$.*

Proof. If F is $\mu_{H'}$-semistable then there is a nontrivial proper saturated subsheaf $F_0 \subset F$ with $\mu_{H'}(F_0) = \mu_{H'}(F)$ and we can choose $H_0 = H'$, and if not the claim follows by lemma 1.4.5. Note that $H \neq H_0$ as F is μ_H-stable. □

Proposition 1.4.7 *Let K be an open chamber of the ample cone, $\overline{K}$ its closure in the ample cone, $H \in K$ and $H' \in \overline{K}$. Then one has*

$$\mu_{H'}\text{-stable} \Rightarrow \mu_H\text{-stable} \Rightarrow \mu_H\text{-semistable} \Rightarrow \mu_{H'}\text{-semistable}$$

for torsion free sheaves of rank r and discriminant Δ on X.

Proof.

1. Let F be μ_H-semistable but not $\mu_{H'}$-semistable. Then by lemma 1.4.5 there is an $H_0 \in [H, H'[\subset K$ and a nontrivial proper saturated subsheaf $F_0 \subset F$ such that F and F_0 are μ_{H_0}-semistable with $\mu_{H_0}(F_0) = \mu_{H_0}(F)$ and $\mu_{H'}(F_0) > \mu_{H'}(F)$. But corollary 1.4.4 yields $\frac{c_1(F_0)}{\operatorname{rk} F_0} = \frac{c_1(F)}{\operatorname{rk} F}$, which contradicts $\mu_{H'}(F_0) > \mu_{H'}(F)$.

2. Let F be $\mu_{H'}$-stable but not μ_H-stable. Then by corollary 1.4.6 there is an $H_0 \in]H', H] \subset K$ and a nontrivial proper saturated subsheaf $F_0 \subset F$ such that F and F_0 are μ_{H_0}-semistable with $\mu_{H_0}(F_0) = \mu_{H_0}(F)$. But corollary 1.4.4 yields $\frac{c_1(F_0)}{\operatorname{rk} F_0} = \frac{c_1(F)}{\operatorname{rk} F}$, which contradicts the $\mu_{H'}$-stability of F. □

Corollary 1.4.8 *Let K be an open chamber of the ample cone, $H, H' \in K$ and F a torsion free sheaf of rank r and discriminant Δ on X. Then F is H-(semi)stable if and only if it is H'-(semi)stable.*

Proof. Let F be H-(semi)stable. Hence it is in particular μ_H-semistable and by proposition 1.4.7 also $\mu_{H'}$-semistable. Let $E \subset F$ be a nontrivial proper saturated subsheaf with $\mu_{H'}(E) = \mu_{H'}(F)$. Then corollary 1.4.4 yields $\mu_H(E) = \mu_H(F)$, and the H-(semi)stability ensures

$$\frac{\chi(E)}{\operatorname{rk} E} (\leq) \frac{\chi(F)}{\operatorname{rk} F}.$$

□

1.4. GENERAL AMPLE DIVISORS

Note that a torsion free sheaf F of rank r and discriminant Δ that is semistable with respect to a general ample divisor H does not need to be H'-semistable when H' is on the boundary of the chamber containing H: there might be a subsheaf $E \subset F$ with $\mu_H(E) < \mu_H(F)$ and $\mu_{H'}(E) = \mu_{H'}(F)$ without satisfying

$$\frac{\chi(E)}{\operatorname{rk} E} \leq \frac{\chi(F)}{\operatorname{rk} F}.$$

Also an H'-stable sheaf does not need to be H-stable. One only has the following weaker result:

Lemma 1.4.9 *Let H' be an ample divisor, H a general ample divisor and F an H'-stable and μ_H-semistable sheaf of rank r and discriminant Δ. Then F is H-stable.*

Proof. Let $E \subset F$ be a proper nontrivial subsheaf with $\mu_H(E) = \mu_H(F)$. By corollary 1.4.4 one has $\frac{c_1(E)}{\operatorname{rk} E} = \frac{c_1(F)}{\operatorname{rk} F}$. Hence $\frac{\chi(E)}{\operatorname{rk} E} < \frac{\chi(F)}{\operatorname{rk} F}$ by the H'-stability of F. □

1.4.2 Walls for one-dimensional sheaves

Even though this case is easier than the case of two-dimensional sheaves, there is a small complication we treat first. Let X be a nonsingular projective surface over an algebraically closed field k of characteristic zero, H an ample divisor on X and F a pure one-dimensional sheaf on X, i.e. all nontrivial subsheaves $E \subset F$ are also one-dimensional.

For $\chi(F) = 0$, the notion of H-(semi)stability is independent of the choice of H. Indeed, using Riemann-Roch one gets that the reduced Hilbert polynomial of F is $p_H(F) = m$, and for a nontrivial subsheaf $E \subset F$ one gets

$$p_H(E) = m + \frac{\chi(E)}{c_1(E).H}.$$

Hence the semistability condition is that $\chi(E) \leq 0$ for all nontrivial proper subsheaves $E \subset F$, with strict inequality in the case of stability. In particular, this prevents one from introducing the notion of a general ample divisor in this particular case. However, we can move away from this case:

Lemma 1.4.10 *A one-dimensional sheaf F on X is H-(semi)stable if and only if $F \otimes H$ is H-(semi)stable.*

Proof. This follows from $p_H(F \otimes H)(m) = p_H(F)(m+1)$ for the reduced Hilbert polynomials and the following lemma 1.4.11. □

Lemma 1.4.11 *Let F be a coherent sheaf on a noetherian scheme X and L a line bundle. Then $\dim(F) = \dim(F \otimes L)$, and F is pure if and only if $F \otimes L$ is pure.*

Proof. For all coherent sheaves E on X the stalks of $(E \otimes L)$ and E are isomorphic everywhere, in particular, one has $\mathrm{Supp}(E \otimes L) = \mathrm{Supp}(E)$ and

$$\dim(E \otimes L) = \dim \mathrm{Supp}(E \otimes L) = \dim \mathrm{Supp}(E) = \dim(E).$$

The second claim follows from the exactness of the functor $\bullet \otimes L$. □

Lemma 1.4.12 *Let F be a pure one-dimensional sheaf and $E \subset F$ a nontrivial subsheaf. Then $0 < c_1(E).H \leq c_1(F).H$. Moreover, if $p_H(E) = p_H(F)$ then $\mathrm{sgn}\chi(E) = \mathrm{sgn}\chi(F)$ and $|\chi(E)| \leq |\chi(F)|$.*

Proof. For any one-dimensional sheaf the first Chern class is effective, and for any effective divisor D one has $D.H > 0$. The first Chern class is additive on exact sequences, hence $c_1(E) = c_1(F) - c_1(F/E)$, and the quotient F/E has dimension at most 1, thus

$$c_1(E).H = c_1(F).H - c_1(F/E).H \leq c_1(F).H. \tag{1.1}$$

If $p_H(E) = p_H(F)$ then $\chi(E) = \chi(F)\frac{c_1(E).H}{c_1(F).H}$, and hence $\mathrm{sgn}\chi(E) = \mathrm{sgn}\chi(F)$ and

$$|\chi(E)| = |\chi(F)|\frac{c_1(E).H}{c_1(F).H} \stackrel{(1.1)}{\leq} |\chi(F)|.$$

□

1.4. GENERAL AMPLE DIVISORS

We restrict to X being a projective K3 surface.

Definition 1.4.13 *Let v be the Mukai vector of a pure one-dimensional sheaf F. For a subsheaf $E \subset F$ we define $L := \chi(E)c_1(F) - \chi(F)c_1(E)$, and for $L \neq 0$ we call*
$$W_L := L^\perp \cap \mathrm{Amp}(X)_{\mathbb{Q}}$$
the v-wall defined by L.

Proposition 1.4.14 *The number of nonempty v-walls is finite for a given Mukai vector $v = (0, v_1, v_2)$ with v_1 effective.*

Proof. It is enough to show that the set $S := \{(c_1(E), \chi(E)) \mid$ there is a subsheaf E of some pure one-dimensional sheaf F with $v(F) = v$ and $p_H(E) = p_H(F)$ for some ample divisor $H\}$ is finite.

As X is a K3 surface, the intersection pairing is nondegenerate, the ample cone is open in $\mathrm{NS}(X)_{\mathbb{Q}}$ and $\mathrm{NS}(X)$ is free. Thus we can choose finitely many ample divisors H that span $\mathrm{NS}(X)_{\mathbb{Q}}$ such that every $D \in \mathrm{NS}(X)$ can be regained from $D.H$.

By lemma 1.4.12 each $(c_1(E), \chi(E)) \in S$ satisfies $0 < c_1(E).A \leq v_1.A$ for any ample divisor A and $|\chi(E)| \leq |v_2|$. In particular there are only finitely many choices for $\chi(E)$ and for $c_1(E).A$ with A any fixed ample divisor, hence one has only finitely many choices for $c_1(E)$. $\square$

Yoshioka proves this fact in [Yos01] section 1.4 for the case $v_1^2 > 0$.

Definition 1.4.15 *An ample $\mathbb{Q}$-divisor H is called v-general if H is not contained in any v-wall. The connected components of the complement of the union of all v-walls are called v-chambers.*

This definition immediately yields:

Lemma 1.4.16 *Let F be a pure sheaf of dimension 1, $E \subset F$ a subsheaf, $L := \chi(E)c_1(F) - \chi(F)c_1(E)$ and H a $v(F)$-general ample $\mathbb{Q}$-divisor. If $L.H = 0$ then one has $L = 0$.*

The following fact is essential in the local analysis of moduli spaces of semistable sheaves on K3 surfaces and explains the importance of a general ample divisor. We include the proof because of the lack of a good reference.

Lemma 1.4.17 *Let $v = (0, v_1, v_2) \in \Lambda(X)$ with v_1 effective and $v_2 \neq 0$, H a v-general ample divisor and F an H-semistable sheaf with Mukai vector v. Then for every nontrivial subsheaf $E \subset F$ with $p_H(E) = p_H(F)$ one has $\chi(E) \neq 0$ and*
$$\frac{v(E)}{\chi(E)} = \frac{v(F)}{\chi(F)}.$$
In particular, if such an F is H-polystable then this holds for every direct summand E of F. Moreover, if v is primitive then an H-semistable sheaf F as above must be already H-stable.

Proof. Let $E \subset F$ with $p_H(E) = p_H(F)$, i.e. $\chi(E)c_1(F).H = \chi(F)c_1(E).H$. By lemma 1.4.16 one has $\chi(E)c_1(F) = \chi(F)c_1(E)$. By assumption $\chi(F) = v_2 \neq 0$, and $c_1(E) \neq 0$ because E is one-dimensional, hence also $\chi(E) \neq 0$, and we get
$$\frac{v(E)}{\chi(E)} = \frac{v(F)}{\chi(F)}.$$
In particular, if F is H-polystable then each direct summand $E \subset F$ is a nontrivial saturated subsheaf of F with $p_H(E) = p_H(F)$.

Let v be primitive, and assume there is a nontrivial saturated proper subsheaf of an H-semistable sheaf F with $p_H(E) = p_H(F)$, so
$$\chi(E)c_1(F) = \chi(F)c_1(E).$$
We write $g := \gcd(|\chi(E)|, |\chi(F)|)$. Thus
$$\frac{\chi(E)}{g}c_1(F) = \frac{\chi(F)}{g}c_1(E).$$
The Euler characteristics satisfy $\chi(F) = \chi(E) + \chi(F/E)$. As the reduced

Hilbert polynomials for E, F and F/E are all the same, i.e.

$$\frac{\chi(E)}{c_1(E).H} = \frac{\chi(F)}{c_1(F).H} = \frac{\chi(F/E)}{c_1(F/E).H},$$

and the denominator is always positive, the Euler characteristics have all the same sign, and one has $|\chi(F)| = |\chi(E)| + |\chi(F/E)| > |\chi(E)|$. Thus

$$1 \leq g \leq |\chi(E)| < |\chi(F)|,$$

and

$$\frac{|\chi(F)|}{g} > 1$$

and this integer divides $c_1(F)$ and $\chi(F)$, which is a contradiction to v being primitive. $\square$

1.5 Semistable sheaves on K3 surfaces

We continue section 1.3 having the notion of a general ample divisor at hand.

Lemma 1.5.1 *Let X be a projective K3 surface, $v = (v_0, v_1, v_2) \in \Lambda(X)$ primitive with either $v_0 > 0$ or $v_0 = 0$, $v_1 \neq 0$ effective and $v_2 \neq 0$, $m \in \mathbb{N}$ and H an mv-general ample divisor. Let E be an H-semistable sheaf with Mukai vector $v(E) = mv$. Then $v^2 \geq -2$. In the case of equality, E is Seshadri equivalent to a sheaf $F^{\oplus m}$ with F the unique (up to isomorphism) H-stable sheaf of Mukai vector $v(F) = v$. In particular, there is no H-stable sheaf with Mukai vector mv for $m \geq 2$ if $v^2 = -2$.*

Proof. As the graded object $gr(E)$ of the Jordan-Hölder filtration is H-polystable with $v(gr(E)) = v(E)$ we can assume that E is H-polystable. Let F be an H-stable direct summand of E. As H is v-general F has Mukai vector nv for some $n \in \mathbb{N}$, and corollary 1.3.5 together with [HL97] corollary 1.2.8 yields $(nv)^2 \geq -2$. Hence $v^2 \geq -\frac{2}{n^2} \geq -2$. In the case of $v^2 = -2$, one has equality everywhere and therefore $n = 1$. F is unique up to isomorphism by proposition 1.3.7. □

1.6 Moduli spaces of sheaves on K3 surfaces

Let X be a projective K3 surface, H an ample divisor on X and $v = (v_0, v_1, v_2) \in \Lambda(X)$. There is a projective coarse moduli space $M_H(v)$ that parametrises H-polystable sheaves with Mukai vector v, see [HL97] chapter 6. Let $M_H^s(v)$ denote the open subset of H-stable sheaves.

Proposition 1.6.1 $M_H^s(v)$ *is nonsingular and each connected component has dimension* $2 + v^2$.

Proof. The original proof is given in Mukai [Muk84], see also [HL97] sections 4.5 and 6.1. □

Theorem 1.6.2 $M_H^s(v)$ *admits a symplectic form.*

Proof. The construction of the symplectic form is originally due to Mukai, see [Muk84]. A detailed proof - for simplicity only for positive rank - is given also in [HL97] chapter 10, the final result being stated in theorem 10.4.3. □

Theorem 1.6.3 *If* $M \subseteq M_H(v)$ *is a connected component with* $M \subseteq M_H^s(v)$ *then one already has* $M = M_H(v)$.

Proof. [KLS06] theorem 4.1. □

The following summary is contained in [KLS06] section 1 - for simplicity only for $v_0 > 0$.

Proposition 1.6.4 *Let* v *be primitive with either* $v_0 > 0$ *or* $v_0 = 0$, $v_1 \neq 0$ *effective and* $v_2 \neq 0$. *Let furthermore* $m \in \mathbb{N}$ *and* H *be* mv-*general.*

1. *If* $M_H(mv)$ *is nonempty then* $v^2 \geq -2$.

1.6. MODULI SPACES OF SHEAVES ON K3 SURFACES

2. Let $v^2 \geq -2$ and $m = 1$. If $v_0 > 0$, v_1 ample or $v^2 \geq 2$ then $M_H^s(v)$ is nonempty.

3. If $v^2 = -2$ then $M_H(mv)$ is empty or consists of a reduced point $[E^{\oplus m}]$ with E an H-stable sheaf of Mukai vector v.

4. Let $v^2 = 0$.

    a) If $M_H^s(mv)$ is nonempty then $M_H(mv) = M_H^s(mv)$ and $M_H(mv)$ is a projective symplectic nonsingular surface.

    b) If $m = 1$ then $M_H(v)$ is a projective K3 surface or empty.

    c) If $m > 1$ and $M_H^s(v) \neq \emptyset$ - e.g. for $v_0 > 0$ or v_1 ample - then $M_H(mv) \cong S^m M_H(v)$.

5. Let $v^2 \geq 2$. Then $M_H(mv)$ is a projective symplectic variety of dimension $2 + m^2 v^2$.

    a) If $m = 1$ then $M_H(v) = M_H^s(v)$, and $M_H(v)$ is nonsingular. If $v_0 > 0$ or v_1 ample then $M_H(v)$ is deformation equivalent to $\text{Hilb}^{\frac{v^2}{2}+1}(X)$.

    b) If $m \geq 2$ then the singular locus of $M_H(mv)$ is nonempty and equals the strictly semistable locus.

        i. If $m = 2$ and $v^2 = 2$ then the singular locus has codimension 2 and $M_H(mv)$ admits a symplectic resolution.

        ii. If $m = 2$ and $v^2 > 2$ or $m > 2$ then $M_H(mv)$ is locally factorial, the singular locus has codimension at least 4 and the singularities are terminal. There is no open neighbourhood of a singular point that admits a symplectic resolution.

In particular, if $M_H^s(mv)$ is nonempty then $M_H(mv)$ and $M_H^s(mv)$ are irreducible.

Proof. Assume first $M_H^s(mv)$ is nonempty. Every connected component of $M_H^s(mv)$ is nonsingular of dimension $2 + m^2 v^2$ by theorem 1.6.1 and carries a symplectic form by theorem 1.6.2. If $m = 1$ then there are no strictly H-semistable sheaves, hence $M_H^s(v) = M_H(v)$. If $M_H^s(mv) = M_H(mv)$ then $M_H(mv)$ is connected by theorem 1.6.3, hence irreducible. In particular, $M_H(v)$ is a projective symplectic nonsingular variety.

1. This holds by lemma 1.5.1.

2. Let $v^2 \geq -2$ and $m = 1$. If $v_0 > 0$ or $v^2 \geq 2$ then $M_H^s(v)$ is nonempty by [KLS06] section 1 and section 2.4. If v_1 is ample then this holds by [Yos01] theorem 8.1.

3. This holds by lemma 1.5.1 as well.

4. Let $v^2 = 0$.

   a) It remains to show that there is no strictly H-semistable sheaf with Mukai vector mv. This follows from Mukai's claim in [Muk87] that $M_H^s(av)$ is nonempty for at most one choice of $a \in \mathbb{Q}$, the proof being somewhat hidden in the paper - see also the proof of [Yos00] lemma 1.8.

   b) Let $m = 1$. Then $M_H^s(v)$ is a K3 surface or empty by [Muk87] theorem 1.4.

   c) Let $m > 1$ and $M_H^s(v) \neq \emptyset$. Then $M_H^s(\tilde{m}v) = \emptyset$ for all $\tilde{m} > 1$ and the canonical morphism $S^m M_H(v) \to M_H(mv)$ is an isomorphism.

5. Let $v^2 \geq 2$.

   a) Let $m = 1$. Then $M_H(v) = M_H^s(v)$ and $M_H(v)$ is nonsingular by the statements at the beginning of the proof. If $v_0 > 0$ or v_1 ample then $M_H(v)$ is deformation equivalent to $\mathrm{Hilb}^{\frac{v^2}{2}+1}(X)$ by [Yos01] theorem 8.1.

   b) The statements for $m \geq 2$ are contained in [KLS06] and [LS06] except that if $m = 2$ and $v^2 > 2$ or $m > 2$ then the singularities are terminal. This is given by [Nam01] corollary 1. □

1.6. MODULI SPACES OF SHEAVES ON K3 SURFACES

The exceptional examples of irreducible symplectic manifolds constructed by O'Grady belong to the case $v^2 = 2$ and $m = 2$: He has chosen $v = (1, 0, -1)$ in [O'G99]. The question whether all symplectic resolutions of all $M_H(mv)$ with $v^2 = 2$, $m = 2$ and H mv-general are irreducible symplectic manifolds is still open.

We want to investigate the cases of moduli spaces not covered by these results.

Chapter 2
Irreducible components

For a nongeneral ample divisor H the moduli space $M_H(v)$ need not be irreducible. In particular, there might occur components containing only strictly semistable sheaves. In this chapter we decompose the moduli space and discuss first results on the existence of symplectic resolutions of components containing no stable sheaves. Our ground field is still $\mathbb{C}$.

2.1 A decomposition

Proposition 2.1.1 *Let X be a projective K3 surface, $v \in \Lambda(X)$, H an ample divisor and M an irreducible component of $M_H(v)$. Then there is a birational projective morphism*
$$g : \prod_{i=1}^{m} S^{n_i} M_i \to M$$
for a suitable decomposition $v = \sum_{i=1}^{m} n_i v_i$ with $n_i \in \mathbb{N}$ and $v_i \in \Lambda(X)$ for $1 \leq i \leq m$ and a suitable choice of pairwise distinct irreducible components $M_i \subset \overline{M_H^s(v_i)}$ for $1 \leq i \leq m$.

Moreover, g induces an isomorphism between $V := \prod_{i=1}^{m} S^{n_i} M_i^s$ and the normalisation of $g(V)_{red}$.

Proof. Let S be the at most countable set of finite tuples $(n_i, M_i)_i$ of pairs of a natural number $n_i \in \mathbb{N}$ and pairwise distinct connected components

$M_i \subset \overline{M_H^s(v_i)}$ for some $v_i \in \Lambda(X)$ such that there is an H-polystable sheaf $\left[\bigoplus_i \bigoplus_{j=1}^{n_i} F_{ij}\right] \in M$ with $F_{ij} \in M_i^s$ for all $1 \leq j \leq n_i$ and all i. For every such tuple $t = (n_i, M_i)_i \in S$ consider the morphism

$$g_t : \prod_i S^{n_i} M_i \to M_H(v), \quad ([F_{ij}]) \mapsto \left[\bigoplus_i \bigoplus_{j=1}^{n_i} F_{ij}\right],$$

whose restriction $g_t|_{\prod_i S^{n_i} M_i^s}$ is injective. Products of irreducible spaces are irreducible, the same is true for images. Hence for all $t \in S$ there is an irreducible component of $M_H(v)$ containing Im g_t, which gives a map σ from S to the set C of irreducible components of $M_H(v)$. One has

$$M^o := M \setminus \bigcup_{c \in C \setminus \{M\}} c \subset \bigcup_{t \in \sigma^{-1}(\{M\})} \text{Im } g_t \subset M,$$

hence

$$\dim M^o \leq \dim \bigcup_{t \in \sigma^{-1}(\{M\})} \text{Im } g_t = \max_{t \in \sigma^{-1}(\{M\})} \dim \text{Im } g_t \leq \dim M.$$

M^o is dense in M, hence $\dim M^o = \dim M$ and there is in particular one $t_0 \in \sigma^{-1}(\{M\})$ such that $\dim \text{Im } g_{t_0} = \dim M$. As both are irreducible and closed, we already have Im $g_{t_0} = M$. The image $U := g_{t_0}(\prod_i S^{n_i} M_i^s)$ is open in M, and g_{t_0} induces isomorphisms

$$g_{t_0}^{-1}(U_{sm}) \xrightarrow{\cong} U_{sm} \quad \text{and} \quad \prod_i S^{n_i} M_i^s \xrightarrow{\cong} \tilde{U}$$

using [Gro61] corollary 4.4.9, where U_{sm} denotes the nonsingular locus of U and $\tilde{U}$ is the normalisation of U_{red}. In particular, g_{t_0} is the claimed birational morphism. $\square$

Therefore we need to understand products and symmetric products.

2.2 Products and symmetric products

For a variety X we denote the singular locus by X^{sing}, and if X is nonsingular then we agree that $\operatorname{codim}_X X^{sing} = \infty$.

Proposition 2.2.1 *1. Let X_i be two symplectic varieties for $i = 1, 2$. Then $X_1 \times X_2$ is a symplectic variety with*

$$\operatorname{codim}_{X_1 \times X_2}(X_1 \times X_2)^{sing} = \min(\operatorname{codim}_{X_1} X_1^{sing}, \operatorname{codim}_{X_2} X_2^{sing}).$$

2. Let X be a symplectic variety and $2 \leq n \in \mathbb{N}$. Then $S^n X$ is a symplectic variety with

$$\operatorname{codim}_{S^n X}(S^n X)^{sing} = \min(\operatorname{codim}_X X^{sing}, \dim X).$$

Moreover, if X is a quasiprojective nonsingular surface then

$$\operatorname{Hilb}^n(X) \to S^n X$$

is a symplectic resolution.

Proof.

1. Beauville lists products of symplectic varieties as examples for symplectic varieties in [Bea00] without a proof, so we give one here. Products of normal varieties are again normal by [Gro65] proposition 6.14.1, and the nonsingular locus of $X_1 \times X_2$ is the product of the nonsingular loci of X_1 and X_2, so

$$\operatorname{codim}_{X_1 \times X_2}(X_1 \times X_2)^{sing} = \min(\operatorname{codim}_{X_1} X_1^{sing}, \operatorname{codim}_{X_2} X_2^{sing}).$$

For $i = 1, 2$ let ω_{X_i} be the symplectic form on the nonsingular locus X_i^{sm} of X_i, $f_i : \tilde{X}_i \to X_i$ a resolution of singularities, $\omega_{\tilde{X}_i}$ the 2-form obtained by extending the pullback of ω_{X_i}, and $p_i : X_1 \times X_2 \to X_i$ and $\tilde{p}_i : \tilde{X}_1 \times \tilde{X}_2 \to \tilde{X}_i$ the canonical projections. Then $(f_1, f_2) : \tilde{X}_1 \times \tilde{X}_2 \to X_1 \times X_2$ is a resolution of singularities,

$$\sum_{i=1}^{2} \left(p_i|_{X_1^{sm} \times X_2^{sm}}\right)^* \omega_{X_i}$$

is a symplectic form on the nonsingular locus $X_1^{sm} \times X_2^{sm}$ of $X_1 \times X_2$, and its pullback extends to the 2-form $\tilde{p}_1^* \omega_{\tilde{X}_1} + \tilde{p}_2^* \omega_{\tilde{X}_2}$ on $\tilde{X}_1 \times \tilde{X}_2$, hence $X_1 \times X_2$ is a symplectic variety.

2. By item 1 X^n is a symplectic variety with

$$\mathrm{codim}_{X^n}(X^n)^{sing} = \mathrm{codim}_X X^{sing}.$$

The symplectic form is invariant under the canonical S_n-action, hence the quotient $S^n X = X^n/S_n$ is a symplectic variety by [Bea00] proposition 2.4. Furthermore, taking the quotient yields a singular locus of codimension $\dim X$, hence

$$\mathrm{codim}_{S^n X}(S^n X)^{sing} = \min(\mathrm{codim}_X X^{sing}, \dim X).$$

Assume that X is a quasiprojective nonsingular surface. [Fog68] theorem 2.4 states that $\mathrm{Hilb}^n(X) \to S^n X$ is a resolution of singularities. By [Bea83] section 6 $\mathrm{Hilb}^n(X) \to S^n X$ is a symplectic resolution if X is projective. But the construction of the symplectic form is local, see also the subsequent remark to [Nak99] theorem 1.10. □

Of course, by definition, products of (nonsingular) symplectic varieties are never irreducible symplectic manifolds.

Lemma 2.2.2 *Let X be a $\mathbb{Q}$-factorial normal variety and G a finite group acting on X such that the geometric quotient X/G exists. Then X/G is $\mathbb{Q}$-factorial.*

Proof. The morphism $X \to X/G$ is finite and surjective and X/G is normal, so by [KM98] lemma 5.16 X/G is $\mathbb{Q}$-factorial. $\square$

Corollary 2.2.3 *Let X_i be normal varieties and $n_i \in \mathbb{N}$ for $i = 1,..,n$. If $\prod_{i=1}^n X_i^{n_i}$ is $\mathbb{Q}$-factorial then $\prod_{i=1}^n S^{n_i} X_i$ is also $\mathbb{Q}$-factorial. This holds in particular if X_i is nonsingular for $i = 1,..,n$.*

Theorem 2.2.4 (Bossière, Serman) *The direct product of two $\mathbb{Q}$-factorial varieties (over $\mathbb{C}$) is again $\mathbb{Q}$-factorial.*

Proof. [BS10]. $\square$

Corollary 2.2.5 *Let X_i be $\mathbb{Q}$-factorial normal varieties and $n_i \in \mathbb{N}$ for $i = 1,..,n$. Then $\prod_{i=1}^n S^{n_i} X_i$ is $\mathbb{Q}$-factorial.*

2.3 Symplectic resolvability of components

Definition 2.3.1 *Let X be a scheme. A (symplectic, $\mathbb{Q}$-factorial, ...) normal quasiprojective variety $\tilde{X}$ with at most terminal singularities together with a proper birational morphism $f : \tilde{X} \to X$ is called a (symplectic, $\mathbb{Q}$-factorial, ...) terminalisation (of X).*

There is an easy criterion available for symplectic varieties:

Proposition 2.3.2 *Let X be a symplectic quasiprojective variety. X has only terminal singularities if and only if the singular locus has codimension at least 4.*

Proof. [Nam01] corollary 1. $\square$

Proposition 2.3.3 *Let X be a singular $\mathbb{Q}$-factorial quasiprojective symplectic variety with $\operatorname{codim}_X X^{sing} \geq 4$. Then X has no symplectic resolution.*

Proof. This is contained in the proof of [KLS06] theorem 6.2. $\square$

Proposition 2.3.4 *Let $g : Y \to X$ be a singular $\mathbb{Q}$-factorial projective symplectic terminalisation of a projective scheme X. Then X admits no projective symplectic resolution.*

Proof. Assume there is a projective symplectic resolution $f : \tilde{X} \to X$. By proposition 1.2.2 the normalisation X' of X_{red} is a projective symplectic variety and f factors through a projective symplectic resolution $h : \tilde{X} \to X'$. Furthermore, g factors through a projective birational morphism $h' : Y \to X'$. As a symplectic variety has trivial canonical divisor, the morphisms h and h' are crepant. This is a contradiction to [Nam06] corollary 1. $\square$

Theorem 2.3.5 *Let X be a projective K3 surface, $v \in \Lambda(X)$, H an ample divisor on X and M an irreducible component of $M_H(v)$. Furthermore, let*

$$g : \prod_{i=1}^{m} S^{n_i} M_i \to M$$

be the projective birational morphism given by proposition 2.1.1, where $M_i \subset \overline{M_H^s(v_i)}$ are pairwise distinct irreducible components.

1. *Assume that for each $1 \leq i \leq m$ there is a $\mathbb{Q}$-factorial projective symplectic terminalisation $\tilde{M}_i \to M_i$.*

   a) *Let $\tilde{M}_i$ be nonsingular for all $1 \leq i \leq m$ and let $v_i^2 \leq 0$ for all $1 \leq i \leq m$ with $n_i > 1$. Then there is a projective symplectic resolution $\tilde{M} \to M$.*

   If $\tilde{M}$ can be chosen to be an irreducible symplectic manifold then it is deformation equivalent to $\tilde{M}_i$ for some $1 \leq i \leq m$ or to a Hilbert scheme of points on a K3 surface.

   b) *Let $\tilde{M}_j$ be singular or let $v_j^2 \geq 2$ and $n_j > 1$ for some j with $1 \leq j \leq m$. Then there is a singular $\mathbb{Q}$-factorial projective symplectic terminalisation $\tilde{M} \to M$. In particular, M admits no projective symplectic resolution.*

2.3. SYMPLECTIC RESOLVABILITY OF COMPONENTS   25

2. Let $U := g(\prod_{i=1}^{m} S^{n_i} M_i^s)$ and U' be the normalisation of U_{red}. Then there is a $\mathbb{Q}$-factorial symplectic terminalisation $\tilde{U} \to U$ and U' is a $\mathbb{Q}$-factorial symplectic variety.

   Moreover, if $v_j^2 \geq 2$ and $n_j > 1$ for some j with $1 \leq j \leq m$ then there is no projective symplectic resolution of M.

Proof.

1. For all $1 \leq i \leq m$ set

$$M_i^{(n_i)} := \begin{cases} \text{Hilb}^{n_i}(\tilde{M}_i) & \text{if } v_i^2 = 0, \\ S^{n_i} \tilde{M}_i & \text{otherwise.} \end{cases}$$

   Recall that M_i consists of exactly one reduced point if $v_i^2 < 0$. One has the sequence

$$\tilde{M} := \prod_{i=1}^{m} M_i^{(n_i)} \to \prod_{i=1}^{m} S^{n_i} \tilde{M}_i \to \prod_{i=1}^{m} S^{n_i} M_i \to M$$

   of projective birational morphisms, and $\tilde{M}$ is a $\mathbb{Q}$-factorial projective symplectic variety with at most terminal singularities by proposition 2.2.1 together with proposition 2.3.2.

   a) Let $\tilde{M}_i$ be nonsingular for all $1 \leq i \leq m$ and let $v_i^2 \leq 0$ for all $1 \leq i \leq m$ with $n_i > 1$. Then $\tilde{M}$ is nonsingular.

   If $M' \to M$ is any other projective birational morphism with M' a projective irreducible symplectic manifold then $\tilde{M}$ and M' are deformation equivalent irreducible symplectic manifolds by theorem 1.1.4. Furthermore, there is at most one j with $1 \leq j \leq m$ and $v_j^2 \geq 0$ and one has $n_j = 1$ or $v_j^2 = 0$ for such a j. In the second case $\tilde{M} = \text{Hilb}^{n_j}(\tilde{M}_j)$, so $\tilde{M}_j$ must be a K3 surface.

   b) Let $\tilde{M}_j$ be singular or let $v_j^2 \geq 2$ and $n_j > 1$ for some j with $1 \leq j \leq m$. Then by proposition 2.2.1 and corollary 2.2.5 $\tilde{M}$ is singular and $\mathbb{Q}$-factorial. Thus proposition 2.3.4 can be applied.

2. Recall that M_i^s is a nonsingular quasiprojective symplectic variety of dimension $2 + v_i^2$ for all $1 \leq i \leq m$. For all $1 \leq i \leq m$ set

$$M_i^{(n_i)} := \begin{cases} \text{Hilb}^{n_i}(M_i^s) & \text{if } v_i^2 = 0, \\ S^{n_i} M_i^s & \text{otherwise.} \end{cases}$$

One has the sequence

$$\tilde{U} := \prod_{i=1}^m M_i^{(n_i)} \to \prod_{i=1}^m S^{n_i} M_i^s \xrightarrow{g} U$$

of projective birational morphisms using propositions 2.1.1 and 2.2.1, and together with corollary 2.2.3 one has that $\tilde{U}$ and $U' \cong \prod_{i=1}^m S^{n_i} M_i^s$ are $\mathbb{Q}$-factorial symplectic varieties, and that $\tilde{U}$ has at most terminal singularities by proposition 2.3.2.

Let $v_j^2 \geq 2$ and $n_j > 1$ for some j with $1 \leq j \leq m$, and assume there is a projective symplectic resolution $f : \tilde{M} \to M$. Then f factors through a projective symplectic resolution of the normalisation M' of M_{red}, and therefore induces a projective symplectic resolution of any open subset of M', hence also of any open subset of $\prod_{i=1}^m S^{n_i} M_i^s$ as well using the isomorphism of proposition 2.1.1. Consider the open subset

$$\prod_{i \neq j} (S^{n_i} M_i^s)_{sm} \times S^{n_j} M_j^s \subset \prod_{i=1}^m S^{n_i} M_i^s,$$

where the index sm denotes taking the nonsingular locus. By propositions 2.2.1 together with corollary 2.2.3 this open subset is a singular $\mathbb{Q}$-factorial symplectic variety with singular locus of codimension at least 4, which has no symplectic resolution by proposition 2.3.3, a contradiction. □

Chapter 3

Moduli spaces for one-dimensional sheaves

3.1 Morphisms between moduli spaces

This section explains a possibility to construct morphisms between moduli spaces. This will yield partial symplectic resolutions of the moduli spaces $M_H(v)$ for a nongeneral ample divisor H and an isomorphism $M_H(0,c,0) \cong M_H(0,c,c.H)$ for any choice of H.

For a category $\mathcal{C}$ let $\mathcal{C}'$ be the functor category with functors $\mathcal{C}^o \to (Sets)$ as objects and natural transformations between functors as morphisms. For an object x of $\mathcal{C}$ let $\underline{x}$ be the functor $y \mapsto \mathrm{Mor}_{\mathcal{C}}(y,x)$.

Definition 3.1.1 *A functor $\mathcal{F} \in \mathrm{Ob}(\mathcal{C}')$ is corepresented by $F \in \mathrm{Ob}(\mathcal{C})$ if there is a $\mathcal{C}'$-morphism $\alpha : \mathcal{F} \to \underline{F}$ such that every morphism $\mu : \mathcal{F} \to \underline{G}$ factors through a unique morphism $\underline{f} : \underline{F} \to \underline{G}$.*

Lemma 3.1.2 *Let $F \in \mathrm{Coh}(S \times X)$ be a flat family of coherent sheaves on a scheme X with parameter scheme S, $p : S \times X \to S$ and $q : S \times X \to X$ the two projections and $L \in \mathrm{Pic}(X)$ a line bundle. Then $F \otimes q^*L$ is a flat family, too.*

Proof. By definition F is a flat family if for all $a \in S \times X$ the stalk F_a is a flat $\mathcal{O}_{S,p(a)}$-module. The claim follows from $(F \otimes q^*L)_a \cong F_a$. □

Proposition 3.1.3 *Let L be a line bundle on a projective K3 surface X with ample divisor H and $v \in \Lambda(X)$. Assume that a sheaf F with Mukai vector v is H-semistable if and only if $F \otimes L$ is H-semistable. Then there is an isomorphism*
$$M_H(v) \cong M_H(v.\mathrm{ch}(L)) \,.$$

Proof. First of all one has $(F \otimes q^*L)_s \cong F_s \otimes L$ for all $s \in S$. By lemma 3.1.2 the assignment $F \mapsto F \otimes q^*L$ yields an invertible natural transformation between the corepresented functors, which in turn yields the claimed isomorphism. □

We are ready to finish the argument we began in section 1.4.2:

Theorem 3.1.4 *Let X be a K3 surface with an ample divisor H and $0 \neq c \in \mathrm{H}^2(X,\mathbb{Z})$ effective. Then there is an isomorphism*
$$M_H(0,c,0) \cong M_H(0,c,c.H) \,,$$
which is induced by tensoring with H, and one has $c.H > 0$.

Proof. Proposition 3.1.3 together with lemma 1.4.10 yields the claimed isomorphism
$$M_H(0,c,0) \cong M_H(0,c,c.H) \,,$$
as
$$(0,c,0).\mathrm{ch}(H) = (0,c,0).(1,H,\frac{1}{2}H^2) = (0,c,c.H) \,,$$
and $c.H > 0$ because $c \neq 0$ is effective and H is ample. □

This justifies why one can assume without loss of generality that $\chi \neq 0$ when investigating the moduli spaces of one-dimensional semistable sheaves on a K3 surface.

3.1. MORPHISMS BETWEEN MODULI SPACES

Lemma 3.1.5 *Let $\mathcal{C}$ be the category of schemes over a field k, and for $i = 1, 2$ let $\mathcal{M}_i \in \mathcal{C}'$ be a functor that is corepresented by $M_i \in \mathcal{C}$ with an open subfunctor $\mathcal{M}_i^s \subset \mathcal{M}_i$ that is corepresented by an open subscheme $M_i^s \subset M_i$. Assume there is a commutative diagram of natural transformations*

$$\begin{array}{ccc} \mathcal{M}_1 & \xrightarrow{\varphi} & \mathcal{M}_2 \\ {\scriptstyle \iota_1}\downarrow & & \downarrow{\scriptstyle \iota_2} \\ \mathcal{M}_1^s & \xleftarrow{\varphi^s} & \mathcal{M}_2^s. \end{array}$$

Then the induced morphisms form the following commutative diagram:

$$\begin{array}{ccc} M_1 & \xrightarrow{f} & M_2 \\ {\scriptstyle i_1}\downarrow & & \downarrow{\scriptstyle i_2} \\ M_1^s & \xleftarrow{f^s} & M_2^s \end{array}$$

Proof. This follows from the uniqueness of the morphism induced by $\iota_2 = \varphi \circ \iota_1 \circ \varphi^s$. $\square$

Proposition 3.1.6 *For $i = 1, 2$ let $\mathcal{M}_i$ the moduli functor of flat families of sheaves with respect to a semistability condition (i) that is corepresented by a projective scheme M_i and $\mathcal{M}_i^s$ the corresponding open subfunctor for (i)-stable sheaves that is corepresented by an open subscheme $M_i^s \subset M_i$. Assume that (2)-stable $\Rightarrow$ (1)-stable $\Rightarrow$ (1)-semistable $\Rightarrow$ (2)-semistable. Then the canonical natural transformations $\varphi : \mathcal{M}_1 \to \mathcal{M}_2$ and $\varphi^s : \mathcal{M}_2^s \to \mathcal{M}_1^s$ yield morphisms $f : M_1 \to M_2$ and $f^s : M_2^s \to M_1^s$, and f induces a projective birational morphism $\overline{f^{-1}(M_2^s)} \to \overline{M_2^s}$. If furthermore M_1 is irreducible then one has $M_1 = \overline{f^{-1}(M_2^s)}$, and $\overline{M_2^s}$ and M_2^s are irreducible.*

Proof. By lemma 3.1.5 there is a commutative diagram

$$\begin{array}{ccc} M_1 & \xrightarrow{f} & M_2 \\ {\scriptstyle i_1}\downarrow & & \downarrow{\scriptstyle i_2} \\ M_1^s & \xleftarrow{f^s} & M_2^s, \end{array}$$

which induces a commutative diagram

$$\begin{array}{ccc} f^{-1}(M_2^s) & \xrightarrow{f'} & M_2^s \\ {\scriptstyle i}\downarrow & & \parallel \\ f^s(M_2^s) & \xleftarrow[f^s]{} & M_2^s. \end{array}$$

By assumption a (1)-semistable sheaf that is (2)-stable is already (1)-stable, i.e. i is surjective. Hence f' is an isomorphism and one has the claimed birational morphism. This morphism is projective as M_1 and M_2 are projective. The last statement is clear. □

The following lemma establishes the assumption of proposition 3.1.6 such that one can construct morphisms between moduli spaces belonging to varying ample divisors. Moreover, one can see that the notion of (semi)stability is independent of the choice of an ample divisor inside a chamber.

Lemma 3.1.7 *Let X be a projective K3 surface, $v = (0, v_1, v_2) \in \Lambda(X)$ with $v_1 \neq 0$ effective and $v_2 \neq 0$, H an ample divisor in a v-chamber K and H' another ample divisor in the closure $\overline{K}$ of K in the ample cone. Then one has H'-stable $\Rightarrow$ H-stable $\Rightarrow$ H-semistable $\Rightarrow$ H'-semistable for sheaves of Mukai vector v.*

Proof. Let F be a sheaf with Mukai vector v, $E \subset F$ a nontrivial proper subsheaf and

$$f : \overline{K} \to \mathbb{Q}, h \mapsto (\chi(E) c_1(F) - \chi(F) c_1(E)).h\,.$$

We consider the two following cases:

1. F is H-semistable, i.e. $f(H) \leq 0$, and we assume, $f(H') > 0$, or

2. F is H'-stable, i.e. $f(H') < 0$, and we assume, $f(H) \geq 0$.

Then there is a $\mathbb{Q}$-divisor $H_0 \in [H, H'[\subset K$ with $f(H_0) = 0$, and lemma 1.4.16 yields the contradiction $f \equiv 0$. □

3.2 Results

Let X be a projective K3 surface, $v = (0, v_1, v_2) \in \Lambda(X)$ primitive with $v_1 \neq 0$ effective, $m \in \mathbb{N}$ and H an ample divisor on X. If $v_2 = 0$ we have seen that one fails to introduce the notion of a general ample divisor but as we showed in theorem 3.1.4 one has an isomorphism

$$M_H(0, mv_1, 0) \cong M_H(0, mv_1, mv_1.H)$$

with $v_1.H > 0$ so that we can assume without loss of generality that $v_2 \neq 0$.

We want to remark that the case of $\overline{M_H^s(mv)}$ with $v^2 = 0$ is not really interesting regarding the locating of new examples for irreducible symplectic manifolds as one knows already all irreducible symplectic manifolds of (complex) dimension two to be K3 surfaces.

Lemma 3.2.1 *Let $v \in \Lambda(X)$ be primitive with $v^2 \geq 0$ and $m \in \mathbb{N}$.*

1. *One has $v^2 = 0$, or $m = 1$, or $v^2 = 2$ and $m = 2$ if and only if mv is primitive or $(mv)^2 \leq 14$, and*

2. *one has $v^2 \geq 2$ and $m > 2$, or $v^2 > 2$ and $m \geq 2$ if and only if mv is not primitive and $(mv)^2 \geq 16$.*

Proof. Let $v^2 = 0$, or $m = 1$, or $v^2 = 2$ and $m = 2$. If $(mv)^2 \geq 16$ then we are in the case $m = 1$, i.e. mv is primitive.

Conversely, let mv be not primitive and $2 \leq (mv)^2 \leq 14$. Then $\frac{v^2}{2}m^2$ is divisible by a square ≥ 2, hence $\frac{v^2}{2}m^2 = 4$ and therefore $m = 2$ and $v^2 = 2$. $\square$

Theorem 3.2.2 *Let X be a projective K3 surface, $v = (0, v_1, v_2) \in \Lambda(X)$ with $v_1 \neq 0$ effective and $v_2 \neq 0$, and H an ample divisor on X. Assume that $M_H^s(v)$ is nonempty. Then one has $v^2 \geq -2$, and in the case of equality, v is primitive and $M_H(v)$ consists of a reduced point. Let now $v^2 \geq 0$.*

1. *Let v be primitive or $v^2 \leq 14$. Then there is a projective symplectic resolution $M \to \overline{M_H^s(v)}$. If H is not v-general then M can be chosen to be a*

symplectic resolution of $M_A(v)$, where A is a v-general ample divisor in a chamber touching H.

Moreover, if M can be chosen to be irreducible symplectic then it is unique up to deformation.

2. Let v be not primitive and $v^2 \geq 16$. Then there is a singular locally factorial (and therefore $\mathbb{Q}$-factorial) projective symplectic terminalisation of $\overline{M_H^s(v)}$, and in particular, there is no projective symplectic resolution of $\overline{M_H^s(v)}$.

Proof. If H is v-general then this is given by proposition 1.6.4. Assume that H is not v-general. The first part is proposition 1.3.7. Assume $v^2 \geq 0$ and let A be a v-general ample divisor in a chamber touching H. Proposition 3.1.6 yields the projective birational morphism $M_A(v) \to \overline{M_H^s(v)}$ using lemma 3.1.7. The other statements are contained in the propositions 1.6.4, 1.1.4 and 2.3.4 using lemma 3.2.1 for the case differentiation. Note that locally factorial varieties are $\mathbb{Q}$-factorial by [Har77] proposition II.6.11. □

Hence there are no new examples of projective irreducible symplectic manifolds lying birationally over the irreducible component $\overline{M_H^s(v)}$ of $M_H(v)$. Furthermore, we have established the assumption of theorem 2.3.5 item 1, which extends the absence result to all irreducible components of $M_H(v)$. The assumption $v_2 \neq 0$ can be omitted due to theorem 3.1.4.

Corollary 3.2.3 *Let X be a projective K3 surface, $v = (0, v_1, v_2) \in \Lambda(X)$ with $v_1 \neq 0$ effective, H an ample divisor on X and M an irreducible component of $M_H(v)$. If there is a projective symplectic resolution $\tilde{M} \to M$ with $\tilde{M}$ an irreducible symplectic manifold then it is deformation equivalent to a symplectic resolution of some $M_A(w)$, where $w \in \Lambda(X)$ and A is some w-general ample divisor.*

Chapter 4
Moduli spaces for (H, A)-semistable sheaves

We introduce the notion of (H, A)-(semi)stability, establish some of its properties and construct a moduli space for (H, A)-semistable sheaves. We assume familiarity with the material presented in [HL97] and use the notation therein.

4.1 Preliminaries

Let X be a noetherian scheme and E a coherent sheaf on X.

Definition 4.1.1 *Let L be a line bundle on X. A section $s \in H^0(X, L)$ is called E-regular if $E \otimes L^\vee \xrightarrow{\cdot s} E$ is injective. A sequence $s_1, ..., s_\ell \in H^0(X, L)$ is E-regular if s_i is $E/(s_1, ..., s_{i-1})(E \otimes L^\vee)$-regular for all $i = 1, ..., \ell$.*

We also say that the divisor $H \in |L|$ is E-regular if the corresponding section $s \in H^0(X, L)$ is E-regular.

Lemma 4.1.2 *Let L and M be two line bundles on X and $s_1, ..., s_\ell \in H^0(X, L)$ an E-regular sequence. Then it is also $E \otimes M$-regular.*

Proof. If $s \in H^0(X, L)$ is E-regular then $E \otimes L^\vee \xrightarrow{\cdot s} E$ is injective. Hence $E \otimes M \otimes L^\vee \xrightarrow{\cdot s} E \otimes M$ is injective as well, i.e. s is $E \otimes M$-regular. Thus if

s_i is $E/(s_1,...,s_{i-1})(E \otimes L^\vee)$-regular, then it is $(E/(s_1,...,s_{i-1})(E \otimes L^\vee)) \otimes M$-regular. The claim now follows because of $(E/(s_1,...,s_{i-1})(E \otimes L^\vee)) \otimes M = (E \otimes M)/(s_1,...,s_{i-1})(E \otimes M \otimes L^\vee)$. □

4.2 Semistable sheaves

Let X be a projective scheme over a field k and H and A two ample line bundles on X. For a coherent sheaf E we write

$$E(mH + nA) := E \otimes H^{\otimes m} \otimes A^{\otimes n}.$$

Recall that the Hilbert polynomial $P_H(E)$ of a coherent sheaf E with respect to H is

$$P_H(E)(m) := \chi(E(mH)),$$

which is a polynomial in m. Indeed:

Lemma 4.2.1 *Let E be a coherent sheaf of dimension d and let $H_1,...,H_d \in |H|$ be an E-regular sequence. Then*

$$P_H(E)(m) = \sum_{i=0}^{d} \chi(E|_{\cap_{j=1}^i H_j}) \binom{m+i-1}{i}.$$

Proof. [HL97] lemma 1.2.1. □

In particular, $P_H(E)$ can be written in the form

$$P_H(E)(m) = \sum_{i=0}^{\dim E} \alpha_i^H(E) \frac{m^i}{i!} \tag{4.1}$$

with $\alpha_i^H(E) \in \mathbb{Q}$.

$\alpha_{\dim E}^H(E)$ is called the multiplicity of E with respect to H.

4.2. SEMISTABLE SHEAVES

Corollary 4.2.2 *Let L be a line bundle. Then $\alpha^H_{\dim E}(E \otimes L)$ is independent of L and always positive for E nontrivial.*

Proof. First note that $\binom{m+i-1}{i} = \frac{(m+i-1)\cdots m}{i!}$ is a polynomial in m of degree i. Hence by lemma 4.2.1 one has

$$\alpha^H_{\dim E}(E \otimes L) = \chi((E \otimes L)|_{\cap_{j=1}^d H_j}).$$

As the sequence $H_1, ..., H_d$ is $E \otimes L$-regular, $(E \otimes L)|_{\cap_{j=1}^d H_j}$ is zerodimensional and

$$\alpha^H_{\dim E}(E \otimes L) = h^0((E \otimes L)|_{\cap_{j=1}^d H_j}) = h^0(E|_{\cap_{j=1}^d H_j}) > 0.$$

$\square$

Definition 4.2.3 *The reduced Hilbert polynomial of a nontrivial coherent sheaf E of dimension d is defined by*

$$p_H(E) := \frac{P_H(E)}{\alpha^H_d(E)}.$$

Notation 4.2.4 *In order to avoid case differentiation for stable and semistable sheaves we follow the notation 1.2.5 in [HL97] using bracketed inequality signs, e.g. an inequality with $(\leq)$ for (semi)stable sheaves means that one has $\leq$ for semistable sheaves and $<$ for stable sheaves.*

Definition 4.2.5 *A coherent sheaf E of dimension d is H-(semi)stable if it is pure and every proper nontrivial subsheaf $F \subset E$ satisfies the condition $p_H(F) (\leq) p_H(E)$.*

For a coherent sheaf E we define

$$P_{H,A}(E)(m,n) := \chi(E(mH + nA)).$$

Clearly one has $P_{H,A}(E)(\bullet, 0) = P_H(E)$.

Lemma 4.2.6 *Let E be a coherent sheaf of dimension d. Then $P_{H,A}(E)(m,n)$ is a polynomial in m and n, and it has degree d in n and m and total degree d.*

Proof. Let $A_1, ..., A_d \in |\mathcal{O}(A)|$ be an E-regular sequence. By lemma 4.2.1 and equation 4.1 one has

$$\begin{aligned} P_{H,A}(E)(m,n) &= P_A(E(mH))(n) \\ &= \sum_{i=0}^{d} \chi(E(mH)|_{\cap_{j=1}^{i} A_j}) \binom{n+i-1}{i} \\ &= \sum_{i=0}^{d} P_H(E|_{\cap_{j=1}^{i} A_j}) \binom{n+i-1}{i} \\ &= \sum_{i=0}^{d} \sum_{k=0}^{d-i} \alpha_k^H(E|_{\cap_{j=1}^{i} A_j}) \frac{m^k}{k!} \binom{n+i-1}{i}. \end{aligned}$$

$\square$

For a nontrivial coherent sheaf E we define

$$p_{H,A}(E)(m,n) := \frac{\chi(E(mH+nA))}{\alpha_{\dim E}^H(E)} \in \mathbb{Q}[m,n].$$

Clearly one has $p_{H,A}(E)(\bullet, 0) = p_H(E)$.

Recall that there is a natural ordering of polynomials in one variable given by the lexicographic ordering of their coefficients. This generalises to polynomials of two variables by the identification $\mathbb{Q}[m,n] = (\mathbb{Q}[m])[n]$, i.e. we consider the elements as polynomials in n and use the ordering of $\mathbb{Q}[m]$ for comparing coefficients.

We introduce another ordering on $\mathbb{Q}[m,n]$ by defining

$$f \leq_0 g \quad :\Leftrightarrow \quad (f(\bullet, 0), -f) \leq (g(\bullet, 0), -g)$$

for $f, g \in \mathbb{Q}[m,n]$, where on the right hand side we use lexicographic ordering on the product $\mathbb{Q}[m] \times \mathbb{Q}[m,n]$, i.e. $f \leq_0 g$ if and only if $f(\bullet, 0) < g(\bullet, 0)$ or $f(\bullet, 0) = g(\bullet, 0)$ and $f \geq g$.

4.2. SEMISTABLE SHEAVES

Clearly one has $f =_0 g$ if and only if $f = g$.

Observation 4.2.7 *This ordering is invariant under rescaling the arguments, i.e. for $f, g \in \mathbb{Q}[m, n]$ and $\tilde{f}(m, n) := f(am, bn)$, $\tilde{g}(m, n) := g(am, bn)$ with $a, b \in \mathbb{Q}^+$ one has $f \leq_0 g$ if and only if $\tilde{f} \leq_0 \tilde{g}$.*

We come to the central notion of this chapter, (H, A)-stability. The definition is motivated by [MW97] as the condition is contained therein without getting a name in the case of X being a surface.

Definition 4.2.8 *A coherent sheaf E of dimension d is (H, A)-(semi)stable if it is pure and if for any proper nontrivial subsheaf $F \subset E$ one has*

$$p_{H,A}(F) \;(\leq_0)\; p_{H,A}(E).$$

If E is strictly (H, A)-semistable, i.e. (H, A)-semistable but not (H, A)-stable, then there is always a proper nontrivial subsheaf $F \subset E$ with $p_{H,A}(F) = p_{H,A}(E)$, which is then called an (H, A)-*destabilising subsheaf*.

By observation 4.2.7 this definition is independent of the choice of the two ample line bundles in $\mathbb{Q} \cdot H \times \mathbb{Q} \cdot A$. In particular, (H, A)-(semi)stability is well-defined for ample $\mathbb{Q}$-line bundles, and H or A can be chosen to be very ample without changing the (H, A)-(semi)stability.

Definition 4.2.8 can be restated as follows: A coherent sheaf E of dimension d is (H, A)-(semi)stable if it is H-semistable and if for any proper nontrivial subsheaf $F \subset E$ with $p_H(F) = p_H(E)$ one has $p_{H,A}(F) \;(\geq)\; p_{H,A}(E)$.

Observation 4.2.9 *One has the implications*

$$H\text{-stable} \Rightarrow (H, A)\text{-stable} \Rightarrow (H, A)\text{-semistable} \Rightarrow H\text{-semistable}.$$

This trivial observation must not be neglected: it is the reason why we can get morphisms between the corresponding moduli spaces. Conversely, there might be (H, A)-stable sheaves that are not H-stable, and there might be H-semistable sheaves that are not (H, A)-semistable.

38 CHAPTER 4. MODULI SPACES FOR (H, A)-SEMISTABLE SHEAVES

(H, A)-(semi)stability is a generalisation of H-(semi)stability in the following sense:

Lemma 4.2.10 (H, H)-*(semi)stability is equivalent to H-(semi)stability, and one has $p_H(F) = p_H(E)$ for two coherent sheaves E and F if and only if $p_{H,H}(F) = p_{H,H}(E)$.*

Proof. Use $p_{H,H}(E)(n, m) = p_H(E)(n+m)$. □

In particular, everything we can prove for (H, A)-(semi)stability also holds for H-(semi)stability.

Conversely, one can generalise known facts on H-(semi)stability to (H, A)-(semi)stability.

Lemma 4.2.11 *Let $0 \to E \to F \to G \to 0$ be an exact sequence of coherent sheaves. Then $P_{H,A}(F) = P_{H,A}(E) + P_{H,A}(G)$. Assume furthermore that all three are of dimension d. Then $\alpha_d^H(F) = \alpha_d^H(E) + \alpha_d^H(G)$. Furthermore, one has*

$$\alpha_d^H(E)(p_{H,A}(E) - p_{H,A}(F)) = \alpha_d^H(G)(p_{H,A}(F) - p_{H,A}(G)).$$

Proof. The functor $\bullet \otimes \mathcal{O}(mH + nA)$ is exact and χ is additive on exact sequences, hence the first two equalities. Let all three sheaves be of dimension d. Then one calculates

$$\begin{aligned}
\alpha_d^H(E)(p_{H,A}(E) - p_{H,A}(F)) &= P_{H,A}(E) - (\alpha_d^H(F) - \alpha_d^H(G))p_{H,A}(F) \\
&= P_{H,A}(E) - P_{H,A}(F) + \alpha_d^H(G)p_{H,A}(F) \\
&= -P_{H,A}(G) + \alpha_d^H(G)p_{H,A}(F) \\
&= \alpha_d^H(G)(p_{H,A}(F) - p_{H,A}(G)).
\end{aligned}$$

□

Lemma 4.2.12 *Let E be an H-semistable sheaf and $F \subset E$ a proper nontrivial subsheaf with $p_H(E) = p_H(F)$. Then F is saturated, i.e. E/F is pure of dimension $\dim E$.*

4.2. SEMISTABLE SHEAVES

Proof. If $d := \dim(E) = 0$ then F is always saturated. Assume $d > 0$ and let F' be the saturation of F in E. Thus one has an exact sequence

$$0 \to F' \to E \to (E/F)/T_{d-1}(E/F) \to 0,$$

see [HL97] section 1.1. Assume F is not saturated, i.e. $T_{d-1}(E/F) \neq 0$. By lemma 4.2.11 one has

$$\begin{aligned} P_H(E) - P_H(F') &= P_H((E/F)/T_{d-1}(E/F)) \\ &= P_H(E) - P_H(F) - P_H(T_{d-1}(E/F)), \end{aligned}$$

hence

$$P_H(F) - P_H(F') = -P_H(T_{d-1}(E/F)).$$

As the right hand side is of degree at most $d-1$, one has $\alpha_d^H(F) = \alpha_d^H(F')$ and thus

$$p_H(F) - p_H(F') = -\frac{P_H(T_{d-1}(E/F))}{\alpha_d^H(F)} < 0.$$

By H-semistability of E this yields the chain

$$p_H(F) < p_H(F') \leq p_H(E)$$

in contradiction to the assumption $p_H(F) = p_H(E)$. □

Corollary 4.2.13 *Let E be an H-semistable sheaf of dimension d and $E \to G$ a proper d-dimensional quotient sheaf with $p_H(E) = p_H(G)$. Then G is pure.*

Proof. Let $F := \ker(E \to G)$. As E is H-semistable, E is in particular pure, thus F is d-dimensional. By lemma 4.2.11 one has $p_H(F) = p_H(E)$ and by lemma 4.2.12 F is saturated, hence $G \cong E/F$ is pure. □

Proposition 4.2.14 *Let E be a pure sheaf of dimension d. Then the following conditions are equivalent:*

1. *E is H-(semi)stable.*

2. For all proper saturated subsheaves $F \subset E$ one has $p_H(F) (\leq) p_H(E)$.

3. For all proper quotient sheaves $E \to G$ with $\alpha_d^H(G) > 0$ one has $p_H(E) (\leq) p_H(G)$.

4. For all proper pure d-dimensional quotient sheaves $E \to G$ one has $p_H(E) (\leq) p_H(G)$.

Proof. [HL97] proposition 1.2.6. $\square$

Proposition 4.2.15 *Let E be an H-semistable sheaf of dimension d. Then the following conditions are equivalent:*

1. E is (H, A)-(semi)stable.

2. For all proper saturated subsheaves $F \subset E$ with $p_H(F) = p_H(E)$ one has $p_{H,A}(F) (\geq) p_{H,A}(E)$.

3. For all proper d-dimensional quotient sheaves $E \to G$ with $p_H(E) = p_H(G)$ one has $p_{H,A}(E) (\geq) p_{H,A}(G)$.

4. For all proper pure d-dimensional quotient sheaves $E \to G$ with $p_H(E) = p_H(G)$ one has $p_{H,A}(E) (\geq) p_{H,A}(G)$.

Proof. The implications 1) $\Rightarrow$ 2) and 3) $\Rightarrow$ 4) are obvious, and the implications 2) $\Rightarrow$ 1) and 4) $\Rightarrow$ 3) are trivial by lemma 4.2.12 and corollary 4.2.13.

Let $0 \to F \to E \to G \to 0$ be an exact sequence with $F \subset E$ proper and nontrivial. Then F is saturated if and only if G is pure and d-dimensional. If one of these conditions and thus both are given then lemma 4.2.11 yields $p_H(F) = p_H(E)$ if and only if $p_H(E) = p_H(G)$ and $p_{H,A}(F) (\geq) p_{H,A}(E)$ if and only if $p_{H,A}(E) (\geq) p_{H,A}(G)$, i.e. 2) $\Leftrightarrow$ 4). $\square$

Corollary 4.2.16 *Let E be a pure sheaf of dimension d. Then the following conditions are equivalent:*

1. E is (H, A)-(semi)stable.

4.2. SEMISTABLE SHEAVES

2. For all proper saturated subsheaves $F \subset E$ one has $p_{H,A}(F)\ (\leq_0)\ p_{H,A}(E)$.

3. For all proper quotient sheaves $E \to G$ with $\alpha_d^H(G) > 0$ one has $p_{H,A}(E)\ (\leq_0)\ p_{H,A}(G)$.

4. For all proper pure d-dimensional quotient sheaves $E \to G$ one has $p_{H,A}(E)\ (\leq_0)\ p_{H,A}(G)$.

Corollary 4.2.17 *Any pure sheaf of rank one is (H, A)-(semi)stable.*

Proof. This follows from the characterisation 2 of corollary 4.2.16. □

Proposition 4.2.18 *Let F and G be two (H, A)-semistable sheaves of dimension d.*

1. *If $p_{H,A}(F) >_0 p_{H,A}(G)$ then $\mathrm{Hom}(F, G) = 0$.*

2. *If $p_{H,A}(F) = p_{H,A}(G)$ and $f : F \to G$ is nontrivial then f is injective if F is (H, A)-stable and surjective if G is (H, A)-stable.*

3. *If $P_{H,A}(F) = P_{H,A}(G)$ then any nontrivial homomorphism $f : F \to G$ is an isomorphism provided F or G is (H, A)-stable.*

Proof. The proof carries over literally from [HL97] proposition 1.2.7. □

Corollary 4.2.19 *If E is an (H, A)-stable sheaf then $\mathrm{End}(E)$ is a finite dimensional division algebra over k. In particular, if k is algebraically closed then $k \cong \mathrm{End}(E)$, i.e. E is a simple sheaf.*

Proof. The proof carries over literally from [HL97] corollary 1.2.8. □

4.3 Jordan-Hölder filtration and S-equivalence

Definition 4.3.1 *Let E be an (H,A)-semistable sheaf of dimension d. A Jordan-Hölder filtration of E is a filtration*

$$0 = E_0 \subset E_1 \subset ... \subset E_\ell = E$$

such that the factors $gr_i(E) := E_i/E_{i-1}$ are (H,A)-stable with $p_{H,A}(gr_i(E)) = p_{H,A}(E)$ for all $i = 1,...,\ell$.

Proposition 4.3.2 *Jordan-Hölder filtrations always exist. The graded object $gr(E) := \bigoplus_{i=1}^{\ell} gr_i(E)$ does not depend on the choice of the Jordan-Hölder filtration.*

Proof. The proof carries over from [HL97] proposition 1.5.2. □

Definition 4.3.3 *Two (H,A)-semistable sheaves E_1 and E_2 with $p_{H,A}(E_1) = p_{H,A}(E_2)$ are called Seshadri equivalent or S-equivalent if $gr(E_1) \cong gr(E_2)$.*

Let E be an (H,A)-semistable sheaf of dimension d and

$$0 = E_0 \subset E_1 \subset ... \subset E_\ell = E$$

a Jordan-Hölder filtration of E. By observation 4.2.9 E is in particular H-semistable but the factors $gr_i(E)$ are not necessarily H-stable. Thus one gets a Jordan-Hölder filtration of E with respect to H-stability by refining the given filtration.

Passing from the set of H-semistable sheaves to the set of (H,A)-semistable sheaves one looses sheaves, and the S-equivalence classes become smaller. This is the reason why a moduli space for (H,A)-semistable sheaves parametrising (H,A)-polystable sheaves can partially resolve a component of a moduli space for H-semistable sheaves parametrising H-polystable sheaves.

Proposition 4.3.4 *If k is algebraically closed and E is an (H,A)-stable sheaf then E is also geometrically (H,A)-stable.*

Proof. The proof carries over from [HL97] section 1.5. □

4.4 Flat families

Proposition 4.4.1 *Let $f : X \to S$ be a projective morphism of noetherian schemes, H and A two f-ample invertible sheaves on X and F a flat family of sheaves on the fibres of f. Then the polynomial $P_{H_s,A_s}(F_s)$ is locally constant as a function of $s \in S$.*

Proof. The family $F(\ell H)$ is S-flat as well for all $\ell \in \mathbb{N}_0$, so by [HL97] proposition 2.1.2 the Hilbert polynomial

$$P_{A_s}(F_s(\ell H_s)) = P_{H_s,A_s}(F_s)(\ell, \bullet) \in \mathbb{Q}[n]$$

is locally constant as a function of $s \in S$ for all $\ell \in \mathbb{N}_0$. The polynomial $P_{H_s,A_s}(F_s)$ can be regained from $P_{H_s,A_s}(F_s)(\ell, \bullet)$ for finitely many choices of ℓ, hence it is locally constant as a function of $s \in S$ as well. □

Proposition 4.4.2 *The following properties of coherent sheaves are open in flat families: being (H, A)-semistable, or (H, A)-stable.*

Proof. Let $f : X \to S$ be a projective morphism of noetherian schemes, H and A two f-very ample invertible sheaves on X and F a flat family of d-dimensional sheaves on the fibres of f with Hilbert polynomial P with respect to H_s for all $s \in S$. As we want to show the openness of certain subsets we can assume S to be connected. Furthermore, we can replace S by the open subset of all $s \in S$ such that F_s is H_s-semistable as this condition is open by [HL97] proposition 2.3.1, having in mind observation 4.2.9. Let $\alpha \in \mathbb{N}$ be the multiplicity associated to P.

For each $\alpha' \in \mathbb{N}$ with $\alpha' \leq \alpha$ we consider the relative Quot scheme

$$\pi : Q(\alpha') := \mathrm{Quot}_{X/S}(F, \frac{\alpha'}{\alpha} P) \to S,$$

see [HL97] section 2.2. Let $C(\alpha')$ be the set of connected components of $Q(\alpha')$ and $U \in \mathrm{Coh}(Q(\alpha') \times_S X)$ the universal quotient family. By proposition 4.4.1 $P_{H,A} := P_{H_s,A_s}(F_s)$ is independent of $s \in S$ and

$$P_{H,A}(C) := P_{H_{\pi(q)},A_{\pi(q)}}(U_q)$$

is independent of $q \in C$ for $C \in C(\alpha')$. Let $p_{H,A}$ and $p_{H,A}(C)$ be the reduced polynomials associated to $P_{H,A}$ and $P_{H,A}(C)$, respectively.

Now F_s is (H_s, A_s)-(semi)stable if and only if it is not contained in the closed union

$$\bigcup_{\alpha'=1}^{\alpha} \pi \left(\bigcup_{C \in C(\alpha')\,:\, p_{H,A}\,(<)\,p_{H,A}(C)} C \right).$$

$\square$

4.5 The moduli functor

We come to the construction of the moduli space of (H, A)-semistable sheaves. We generalise the construction in [HL97] chapter 4 according to the idea in [MW97]. There are only very little changes, so it might be very repetitive for readers familiar with the book [HL97].

Let X be a projective scheme over an algebraically closed field k with two ample line bundles H and A. For a fixed polynomial $P \in \mathbb{Q}[m]$ define a functor

$$\mathcal{M}' : (Sch/k)^o \to (Sets)$$

from the category opposed to the category of k-schemes to the category of sets as follows. For a k-scheme S let $\mathcal{M}'(S)$ be the set of all isomorphism classes of S-flat families of (H, A)-semistable sheaves on X with Hilbert polynomial P, and for a k-morphism $f : S' \to S$ let

$$\mathcal{M}'(f) : \mathcal{M}'(S) \to \mathcal{M}'(S'), [F] \mapsto [(f \times \mathrm{id}_X)^* F].$$

If we consider the equivalence relation $F \sim F'$ for two $F, F' \in \mathcal{M}'(S)$ if and only if $F \cong F' \otimes p^*L$ for some $L \in \text{Pic}(S)$, where $p : S \times X \to S$ is the projection onto the first factor, then we get our moduli space functor as quotient functor:

$$\mathcal{M} := \mathcal{M}'/\sim$$

is the moduli functor for (H, A)-semistable sheaves on X with Hilbert polynomial P.

Considering only families of (H, A)-stable sheaves yields open subfunctors $\mathcal{M}'^s \subset \mathcal{M}'$ and $\mathcal{M}^s \subset \mathcal{M}$ as the stability condition is open in flat families, see proposition 4.4.2. Note that as our ground field is algebraically closed, being (H, A)-stable is equivalent to being geometrically (H, A)-stable, see lemma 4.3.4.

Definition 4.5.1 *A scheme M is called a moduli space for (H, A)-semistable sheaves if M corepresents $\mathcal{M}$. We will denote M by $M_{H,A}(P)$, and analogously the functors.*

Lemma 4.5.2 *Suppose M corepresents $\mathcal{M}$. Then Seshadri equivalent sheaves correspond to identical closed points in M. In particular, if there is a properly (H, A)-semistable sheaf F, then $\mathcal{M}$ cannot be represented.*

Proof. The proof carries over literally from [HL97] lemma 4.1.2. $\square$

4.6 The construction of the moduli space

Let X be a connected projective scheme over an algebraically closed field k of characteristic zero, H and A two ample line bundles on X and $P \in \mathbb{Q}[x]$.

At first we follow exactly the construction in [HL97] section 4.3. Details can be found therein. According to [HL97] theorem 3.3.7 the family of H-semistable sheaves on X with Hilbert polynomial with respect to H equal to P is bounded. In particular, there is an integer m such that any such sheaf

F is m-regular. Let $V := k^{\oplus P(m)}$ and $\mathcal{H} := V \otimes_k \mathcal{O}_X(-mH)$. Then there is a surjection
$$\rho : \mathcal{H} \to F,$$
which gives a closed point
$$[\rho : \mathcal{H} \to F] \in R \subset \operatorname{Quot}(\mathcal{H}, P),$$
where $\operatorname{Quot}(\mathcal{H}, P)$ is Grothendieck's Quot scheme of quotients of $\mathcal{H}$ with Hilbert polynomial P on X, see e.g. [HL97] section 2.2, and R is the open subset of $\operatorname{Quot}(\mathcal{H}, P)$ of all quotients $[\mathcal{H} \to E]$, where E is H-semistable and the induced map
$$V = H^0(\mathcal{H}(mH)) \to H^0(E(mH))$$
is an isomorphism.

Let $R^{ss} \subset R$ denote the open subscheme of those points which parametrise (H, A)-semistable sheaves, and $R^s \subset R$ the open subscheme of those parametrising (H, A)-stable sheaves.

There is a $\operatorname{Gl}(V)$-action on $\operatorname{Quot}(\mathcal{H}, P)$, and R, R^{ss} and R^s are $\operatorname{Gl}(V)$-invariant.

Lemma 4.6.1 *If $R^{ss} \to M$ is a categorical quotient for the $\operatorname{Gl}(V)$-action then M corepresents the functor $\mathcal{M}'$. Conversely, if M corepresents $\mathcal{M}'$ then the morphism $R^{ss} \to M$ induced by the universal quotient module on $R^{ss} \times X$ is a categorical quotient. Similarly, $R^s \to M^s$ is a categorical quotient if and only if M^s corepresents the functor $\mathcal{M}^s$.*

Proof. The proof from [HL97] lemma 4.3.1 carries over literally. □

Lemma 4.6.2 *Let $[\rho : \mathcal{H} \to F] \in \operatorname{Quot}(\mathcal{H}, P)$ be a closed point such that $F(mH)$ is globally generated and such that the induced map $H^0(\rho(mH)) : H^0(\mathcal{H}(mH)) \to H^0(F(mH))$ is an isomorphism. Then there is a natural injective homomorphism $\operatorname{Aut}(F) \to \operatorname{Gl}(V)$ whose image is precisely the stabiliser subgroup $\operatorname{Gl}(V)_{[\rho]}$ of the point $[\rho]$.*

4.6. THE CONSTRUCTION OF THE MODULI SPACE

Proof. [HL97] lemma 4.3.2. □

Proposition 4.6.3 *If ℓ is sufficiently large then the line bundle*

$$L_\ell := \det(p_*(\mathcal{F} \otimes q^*\mathcal{O}_X(mH + \ell A)))$$

on $\mathrm{Quot}(\mathcal{H}, P)$ *is very ample and carries a natural* $\mathrm{Gl}(V)$-*linearisation, where p and q are the two projections from* $\mathrm{Quot}(\mathcal{H}, P) \times X$ *to the first and second factor, respectively, and $\mathcal{F}$ is the universal quotient sheaf on* $\mathrm{Quot}(\mathcal{H}, P) \times X$, *see [HL97] section 2.2.*

Proof. If E is a sheaf with Hilbert polynomial P with respect to H then $E(mH)$ has Hilbert polynomial

$$P_H(E(mH))(x) = P(x + m) =: P'(x)$$

with respect to H. As tensoring with a line bundle is exact, one has an isomorphism

$$\varphi : \mathrm{Quot}(\mathcal{H}, P) \to \mathrm{Quot}(\mathcal{H}(mH), P')$$

and

$$L'_\ell := (\varphi^{-1})^* L_\ell = \det(p'_*(\mathcal{F}' \otimes q'^*\mathcal{O}_X(\ell A))),$$

where p' and q' are the two projections from $\mathrm{Quot}(\mathcal{H}(mH), P') \times X$ to the first and second factor, respectively, and $\mathcal{F}'$ is the universal quotient sheaf on $\mathrm{Quot}(\mathcal{H}(mH), P') \times X$. So we can assume without loss of generality that $m = 0$.

Let $S \subset \mathrm{Quot}(\mathcal{H}, P)$ be a connected component. The universal family $\mathcal{F}$ is $\mathrm{Quot}(\mathcal{H}, P)$-flat, hence the Hilbert polynomials $P_A(\mathcal{F}_s)$ are constant on S by [HL97] proposition 2.1.2, say $P_A(\mathcal{F}_s) = P'$ for all $s \in S$. Hence one has an injective morphism

$$\psi : S \to \mathrm{Quot}_A(\mathcal{H}, P'),$$

where the index A denotes that the Hilbert polynomial is with respect to A, and not to H as before. For sufficiently large ℓ

$$L'_\ell := \det(p'_*(\mathcal{F}' \otimes q'^* \mathcal{O}_X(\ell A)))$$

is very ample by proposition [HL97] 2.2.5, where p' and q' are the two projections from $\mathrm{Quot}_A(\mathcal{H}, P') \times X$ to the first and second factor, respectively, and $\mathcal{F}'$ is the universal quotient sheaf on $\mathrm{Quot}_A(\mathcal{H}, P') \times X$, and L'_ℓ carries a natural $\mathrm{Gl}(V)$-linearisation as explained in [HL97] section 4.3. Thus $L_\ell|_S = \psi^* L'_\ell$ is very ample as well and carries a natural $\mathrm{Gl}(V)$-linearisation, hence also L_ℓ itself. $\square$

As the center of $\mathrm{Gl}(V)$ is contained in the stabiliser of each point in $\mathrm{Quot}(\mathcal{H}, P)$ we can restrict the action to $\mathrm{Sl}(V)$. Thus one has the notion of (semi)stable points of $\mathrm{Quot}(\mathcal{H}, P)$ with respect to L_ℓ and the $\mathrm{Sl}(V)$-action.

Parts of the following theorem are contained in [MW97] Key GIT lemma 2.4 but only for X being a surface.

Theorem 4.6.4 *Suppose that m, and for fixed m also ℓ are sufficiently large integers. Then $R^{ss} = \overline{R}^{ss}(L_\ell)$ and $R^s = \overline{R}^s(L_\ell)$. Moreover, the closures of the orbits of two points $[\rho_i : \mathcal{H} \to F_i]$, $i = 1, 2$, in R^{ss} intersect if and only if $\mathrm{gr}^{JH}(F_1) \cong \mathrm{gr}^{JH}(F_2)$. The orbit of a point $[\rho : \mathcal{H} \to F]$ is closed in R^{ss} if and only if F is polystable.*

The proof of this theorem will take up section 4.7. Together with lemma 4.6.1 and [HL97] theorem 4.2.10 it yields:

Theorem 4.6.5 *There is a projective scheme $M_{H,A}(P)$ that universally corepresents the functor $\mathcal{M}_{H,A}(P)$. Closed points in $M_{H,A}(P)$ are in bijection with Seshadri equivalence classes of (H, A)-semistable sheaves with Hilbert polynomial P. Moreover, there is an open subset $M^s_{H,A}(P)$ that universally corepresents the functor $\mathcal{M}^s_{H,A}(P)$.*

4.7 The construction - Proofs

Theorem 4.7.1 *Let p be a polynomial of degree d, and let r be a positive integer. Then for all sufficiently large integers m the following properties are equivalent for a purely d-dimensional sheaf F of multiplicity r and reduced Hilbert polynomial p with respect to H.*

1. *F is H-(semi)stable.*

2. $rp(m) \leq h^0(F(mH))$, *and* $h^0(F'(mH)) (\leq) r'p(m)$ *for all subsheaves* $F' \subset F$ *of multiplicity* r', $0 < r' < r$.

3. $r''p(m) (\leq) h^0(F''(mH))$ *for all quotient sheaves* $F \to F''$ *of multiplicity* r'', $r > r'' > 0$.

Moreover, for sufficiently large m, equality holds in 2. and 3. if and only if F' or F'', respectively, are destabilising.

Proof. [HL97] theorem 4.4.1. □

Proposition 4.7.2 *If F is a coherent $\mathcal{O}_X$-module of dimension d which can be deformed to a pure sheaf, then there exists a pure sheaf E with $P_H(E) = P_H(F)$ and a homomorphism $\varphi : F \to E$ with $\ker \varphi = T_{d-1}(F)$.*

Proof. [HL97] proposition 4.4.2. □

Let $[\rho : V \otimes \mathcal{O}_X(-mH) \to F]$ be a closed point in $\overline{R}$, $\lambda : \mathbb{G}_m \to \mathrm{Sl}(V)$ a one-parameter subgroup and $V = \bigoplus_{n \in \mathbb{Z}} V_n$ the weight space decomposition. Define ascending filtrations on V and F by

$$V_{\leq n} := \bigoplus_{\nu \leq n} V_\nu \quad \text{and} \quad F_{\leq n} := \rho(V_{\leq n} \otimes \mathcal{O}_X(-mH)).$$

Then ρ induces surjections $\rho_n : V_n \otimes \mathcal{O}_X(-mH) \to F_n := F_{\leq n}/F_{\leq n-1}$. Summing up over all weights we get a closed point

$$\left[\overline{\rho} := \bigoplus_{n \in \mathbb{Z}} \rho_n : V \otimes \mathcal{O}_X(-mH) \to \overline{F} := \bigoplus_{n \in \mathbb{Z}} F_n\right] \in \mathrm{Quot}(\mathcal{H}, P).$$

50 CHAPTER 4. MODULI SPACES FOR (H,A)-SEMISTABLE SHEAVES

Lemma 4.7.3 $[\bar{\rho}] = \lim_{t \to 0}[\rho] \cdot \lambda(t)$.

Proof. [HL97] lemma 4.4.3. □

Lemma 4.7.4 *The weight of the action of $\mathbb{G}_m$ via λ on the fibre of L_ℓ at the point $[\bar{\rho}]$ is given by*

$$\sum_{n \in \mathbb{Z}} n\chi(F_n(mH + \ell A)) =$$

$$-\frac{1}{\dim(V)} \sum_{n \in \mathbb{Z}} (\dim(V)\chi(F_{\leq n}(mH + \ell A)) - \dim(V_{\leq n})\chi(F(mH + \ell A))) .$$

Proof. This is [HL97] lemma 4.4.4 with minor changes due to the more general situation. However the proof is the same:

$\mathbb{G}_m$ acts on the direct summands F_n of $\overline{F}$ via a character of weight n, hence it acts with weight n on the complex which defines the cohomology groups $H^i(F_n(mH + \ell A))$, $i \geq 0$. This complex has (virtual) total dimension

$$\sum_i (-1)^i h^i(F_n(mH + \ell A)) = \chi(F_n(mH + \ell A)),$$

so that $\mathbb{G}_m$ acts on the determinant with weight $n\chi(F_n(mH + \ell A))$. Since

$$L_\ell([\bar{\rho}]) = \bigotimes_{n \in \mathbb{Z}} \det(H^*(F_n(mH + \ell A))),$$

the weight of the action of $\mathbb{G}_m$ via λ on $L_\ell([\bar{\rho}])$ is $\sum_{n \in \mathbb{Z}} n\chi(F_n(mH + \ell A))$. This can be rewritten in the claimed form, see also [HL97] section 4.4. □

Lemma 4.7.5 *A closed point $[\rho : \mathcal{H} \to F] \in \overline{R}$ is (semi)stable if and only if for all nontrivial proper linear subspaces $V' \subset V$ and the induced sheaf $F' := \rho(V' \otimes \mathcal{O}_X(-mH)) \subset F$ the following inequality holds:*

$$\dim V \cdot \chi(F'(mH + \ell A)) \; (\geq) \; \dim(V') \cdot \chi(F(mH + \ell A)).$$

4.7. THE CONSTRUCTION - PROOFS

Proof. This is the generalisation of [HL97] lemma 4.4.5. The proof carries over literally using the replacement

$$P(\bullet, \ell) \mapsto \chi(\bullet(mH + \ell A)).$$

□

In the following we denote $H^0(\rho(mH))^{-1}(H^0(F'(mH)))$ by $V \cap H^0(F'(mH))$.

Lemma 4.7.6 *If ℓ is sufficiently large, a closed point $[\rho : \mathcal{H} \to F] \in \overline{R}$ is (semi)stable if and only if for all coherent subsheaves $F' \subset F$ and $V' = V \cap H^0(F'(mH))$ the following inequality holds:*

$$\dim V \cdot \chi(F'(mH + zA)) \;(\geq)\; \dim(V') \cdot \chi(F(mH + zA))$$

as polynomials in z.

Proof. This is the generalisation of [HL97] lemma 4.4.6. The proof carries over almost literally again. □

Recall our choice of the ordering $\leq$ on $\mathbb{Q}[m,n]$ explained in section 4.2.

Lemma 4.7.7 *Let $M \subset \mathbb{Q}[m,n]$ be a finite set of polynomials. Then there is an $m_0 \in \mathbb{N}$ such that for all $m' \geq m_0$ and for all $P, Q \in M$ the following conditions are equivalent:*

1. $P \leq Q$,

2. $P(m', \bullet) \leq Q(m', \bullet)$ *(as polynomials in n)* and

3. $P(m', n') \leq Q(m', n')$ *for some $n' \gg 0$.*

Proof. If $|M| = 1$ then all three conditions are always satisfied for any choice made, so one can assume $|M| \geq 2$. Let $P, Q \in M$ with $P \neq Q$. Expand $P - Q = \sum_{i=0}^{d} a_i n^i$ with $d = \deg_n(P - Q)$ and $a_i \in \mathbb{Q}[m]$ for $i = 0, .., d$ and choose $m_{P,Q} \in \mathbb{N}$ large enough such that $a_d < 0$ if and only if $a_d(m') < 0$ for all $m' \geq m_{P,Q}$. Let $m_0 := \max\{m_{P,Q} \mid P, Q \in M \text{ with } P \neq Q\}$. This one does the job: Choose $m' \geq m_0$ and $P, Q \in M$. If $P = Q$ then $P(m', \bullet) = Q(m', \bullet)$. If $P < Q$ then $P(m', \bullet) < Q(m', \bullet)$ because $m' \geq m_0 \geq m_{P,Q}$. Hence condition 1 is equivalent to condition 2. The equivalence of condition 2 and condition 3 is clear. $\square$

Proposition 4.7.8 *Let $P \in \mathbb{Q}[m]$ be a polynomial. The set $M := \{P_{H,A}(G) \mid G \text{ is a quotient of an } H\text{-semistable sheaf } F \text{ with } P_H(F) = P \text{ and } p_H(F) = p_H(G)\} \subset \mathbb{Q}[m, n]$ is finite.*

Proof. An immediate consequence of the definition of H-semistability is that any H-destabilising quotient of an H-semistable sheaf is H-semistable as well. Let α be the multiplicity associated to P, $\alpha' \in \mathbb{N}$ with $\alpha' \leq \alpha$ and $\mathcal{F} = (F_i)_{i \in I}$ the family of H-semistable sheaves with Hilbert polynomial $\alpha' p$ with respect to H. By [HL97] theorem 3.3.7 this family is bounded. Hence the family $\mathcal{F}(m'H) := (F_i(m'H))_{i \in I}$ is bounded as well for any choice of $m' \in \mathbb{N}$. Therefore by [HL97] lemma 1.7.6 the set of Hilbert polynomials $\{P_A(F_i(m'H)) \mid i \in I\}$ is finite for any choice of $m' \in \mathbb{N}_0$. As the polynomials $P_{H,A}(F_i)$ can be regained from $P_A(F_i(m'H))$ for finitely many choices of m' the set $M_{\alpha'} := \{P_{H,A}(F_i) \mid i \in I\}$ is finite. Altogether one has that M is finite because $M \subset \bigcup_{\alpha'=1}^{\alpha} M_{\alpha'}$. $\square$

Corollary 4.7.9 *Let P be a polynomial in one variable. Suppose that m, and for fixed m also ℓ are sufficiently large integers. Then the following conditions for an H-semistable sheaf F with $P_H(F) = P$ are equivalent:*

1. *F is (H, A)-(semi)stable,*

2. *for all nontrivial proper subsheaves $F' \subset F$ with $p_H(F') = p_H(F)$ one has $p_{H,A}(F') (\geq) p_{H,A}(F)$,*

3. *for all nontrivial proper subsheaves $F' \subset F$ with $p_H(F') = p_H(F)$ one has*

$$\frac{\chi(F'(mH + zA))}{r'} (\geq) \frac{\chi(F(mH + zA))}{r}$$

4.7. THE CONSTRUCTION - PROOFS

as polynomials in z, where r' and r denotes the multiplicity of the sheaves F' and F, and

4. for all nontrivial proper subsheaves $F' \subset F$ with $p_H(F') = p_H(F)$ one has

$$\frac{\chi(F'(mH + \ell A))}{r'} (\geq) \frac{\chi(F(mH + \ell A))}{r}.$$

Proof. The equivalence of conditions 1 and 2 follows immediately from the definition, see also section 4.2. The equivalence of conditions 2-4 is established using lemma 4.7.7 and proposition 4.7.8. □

Proof of theorem 4.6.4. Let m be large enough in the sense of theorem 4.7.1 and of corollary 4.7.9 and such that any H-semistable sheaf with multiplicity $\rho \leq r$ and Hilbert polynomial $\rho \cdot p$ with respect to H is m-regular. Moreover, let ℓ be large enough in the sense of lemma 4.7.6, proposition 4.6.3 and corollary 4.7.9.

First assume that $[\rho : \mathcal{H} \to F]$ is a closed point in R. By definition of R, the map $V \to H^0(F(mH))$ is an isomorphism. Let $F' \subset F$ be a subsheaf of multiplicity $0 < r' < r$ and let $V' = V \cap H^0(F'(mH))$. According to theorem 4.7.1 one has either

1. $p_H(F') = p_H(F)$, or

2. $h^0(F'(mH)) < r' \cdot p(m)$.

In the first case F' is m-regular, and we get $\dim(V') = h^0(F'(mH)) = r' \cdot p(m)$ and therefore

$$\frac{1}{rr'} \left(\dim V \cdot \chi(F'(mH + \ell A)) - \dim(V') \cdot \chi(F(mH + \ell A)) \right)$$
$$= \frac{\dim V}{r} \cdot \frac{\chi(F'(mH + \ell A))}{r'} - \frac{\dim(V')}{r'} \cdot \frac{\chi(F(mH + \ell A))}{r}$$
$$= \frac{\dim V}{r} \left(\frac{\chi(F'(mH + \ell A))}{r'} - \frac{\chi(F(mH + \ell A))}{r} \right).$$

In the second case

$$\dim(V) \cdot r' = rr'p(m) > h^0(F'(mH)) \cdot r = \dim(V') \cdot r.$$

These are the leading coefficients of the polynomials of lemma 4.7.6 up to a factor, so that

$$\dim V \cdot \chi(F'(mH + zA)) > \dim(V') \cdot \chi(F(mH + zA))$$

as polynomials in z. By lemma 4.7.6 and corollary 4.7.9 this proves the implications

1. $[\rho] \in R^s \Rightarrow [\rho] \in \overline{R}^s(L_\ell)$,
2. $[\rho] \in R^{ss} \setminus R^s \Rightarrow [\rho] \in \overline{R}^{ss}(L_\ell) \setminus \overline{R}^s(L_\ell)$ and
3. $[\rho] \in R \setminus R^{ss} \Rightarrow [\rho] \notin \overline{R}^{ss}(L_\ell)$.

Conversely, suppose that $[\rho : V \otimes \mathcal{O}_X(-mH) \to F] \in \overline{R}^{ss}(L_\ell)$. It remains to show that $[\rho] \in R$. By lemma 4.7.6 one has an inequality

$$\dim V \cdot \chi(F'(mH + zA)) \geq \dim(V') \cdot \chi(F(mH + zA))$$

as polynomials in z for any $F' \subset F$ and $V' = V \cap H^0(F'(mH))$. Passing to the leading coefficient of the polynomials we get

$$p(m) \cdot r \cdot r' = \dim(V) \cdot r' \geq \dim(V') \cdot r.$$

This is the inequality (4.4) in the proof of [HL97] theorem 4.3.3 in chapter 4.4, and the remaining part of this proof carries over literally. □

4.8 Local properties and dimension estimates

Let X be a connected projective scheme over an algebraically closed field k of characteristic zero, H and A two ample line bundles on X and $P \in \mathbb{Q}[x]$.

4.9. UNIVERSAL FAMILIES

Theorem 4.8.1 *Let F be an (H, A)-stable sheaf on X represented by a point $[F] \in M_{H,A}(P)$. Then the completion of the local ring $\mathcal{O}_{M_{H,A}(P),[F]}$ pro-represents the deformation functor $\mathcal{D}_F$.*

Proof. The proof of [HL97] theorem 4.5.1 carries over literally. □

As a consequence of this theorem and [HL97] proposition 2.A.11 we get

Corollary 4.8.2 *Let $[F]$ be a stable point. Then the Zariski tangent space of $M_{H,A}(P)$ at $[F]$ is canonically given by $T_{[F]}M_{H,A}(P) \cong \mathrm{Ext}^1(F,F)$. If $\mathrm{Ext}^2(F,F) = 0$ then $M_{H,A}(P)$ is nonsingular at $[F]$. In general there are bounds*

$$\mathrm{ext}^1(F,F) \geq \dim_{[F]} M_{H,A}(P) \geq \mathrm{ext}^1(F,F) - \mathrm{ext}^2(F,F).$$

Theorem 4.8.3 *Let X be a nonsingular projective variety and F an (H,A)-stable sheaf of rank $r > 0$ and determinant bundle Q. Let $M(Q)$ be the fibre of the morphism $\det : M_{H,A}(P) \to \mathrm{Pic}(X)$ over the point $[Q]$. Then $T_{[F]}M(Q) \cong \mathrm{Ext}^1(F,F)_0$. If $\mathrm{Ext}^2(F,F)_0 = 0$ then $M_{H,A}(P)$ and $M(Q)$ are nonsingular at $[F]$. Moreover,*

$$\mathrm{ext}^1(F,F)_0 \geq \dim_{[F]} M(Q) \geq \mathrm{ext}^1(F,F)_0 - \mathrm{ext}^2(F,F)_0.$$

Proof. See [HL97] theorem 4.5.4. □

4.9 Universal families

Definition 4.9.1 *A flat family E of (H, A)-stable sheaves on X parametrised by $M_{H,A}^s(P)$ is called universal if the following holds: if F is an S-flat family of (H, A)-stable sheaves on X with Hilbert polynomial P and if $\Phi_F : S \to M_{H,A}^s(P)$ is the induced morphism then there is a line bundle L on S such that $F \otimes p^*L \cong \Phi_F^* E$, where $p : S \times X \to S$ is the projection. An $M_{H,A}^s(P)$-flat family E is called quasi-universal if there is a locally free $\mathcal{O}_S$-module W such that $F \otimes p^*W \cong \Phi_F^* E$.*

Proposition 4.9.2 *There exist* $\mathrm{Gl}(V)$-*linearised vector bundles on* R^s *with* Z-*weight 1, where* $Z \subset \mathrm{Gl}(V)$ *denotes the center. If* A *is any such vector bundle then* $\mathcal{H}om(p^*A, \tilde{F})$ *descends to a quasi-universal family* E*, where* $\tilde{F}$ *is the universal quotient restricted to* R^s*, and any quasi-universal family arises in this way. If* A *is a line bundle then* E *is universal.*

Proof. The proof carries over from [HL97] proposition 4.6.2 literally. Recall that by corollary 4.2.19 R^s parametrises simple sheaves, i.e. lemma [HL97] 4.6.3 holds as well. □

Chapter 5

Moduli spaces for two-dimensional sheaves

In this chapter we investigate the moduli spaces of torsion free semistable sheaves on a K3 surface. The main question is still whether there is a symplectic resolution of the moduli spaces $M_H(v)$ for a not v-general ample divisor H. We follow the idea of constructing a $\mathbb{Q}$-factorial symplectic terminalisation. A good candidate will be the moduli space for (H, A)-semistable sheaves with v-general A.

5.1 Semistable sheaves

Let X be a nonsingular projective surface over an algebraically closed field k of characteristic zero, H and A two ample divisors on X, K_X the canonical divisor of X and E a coherent sheaf on X. For a line bundle L Riemann-Roch yields

$$\chi(E \otimes L) = \frac{\operatorname{rk} E}{2} c_1(L)^2 + \left(c_1(E) - \frac{\operatorname{rk} E}{2} K_X \right). c_1(L) + \chi(E),$$

and therefore

$$\begin{aligned}
P_{H,A}(E)(m,n) &= \frac{\operatorname{rk} E}{2}(mH+nA)^2 + \left(c_1(E) - \frac{\operatorname{rk} E}{2}K_X\right).(mH+nA) + \chi(E) \\
&= P_H(E)(m) + \alpha_2^H(E)\left(n^2\frac{A^2}{H^2} + 2mn\frac{H.A}{H^2}\right) + \alpha_1^A(E)n.
\end{aligned}$$

In particular, $P_{H,A}(E)$ depends only on $P_H(E)$ and $P_A(E)$. If $\operatorname{rk} E > 0$ then

$$H^2 \cdot p_{H,A}(E)(m,n) = \frac{1}{2}(mH+nA)^2 + \left(\frac{c_1(E)}{\operatorname{rk} E} - \frac{1}{2}K_X\right).(mH+nA) + \frac{\chi(E)}{\operatorname{rk} E}$$

and

$$H^2 \cdot p_H(E)(m) = \frac{1}{2}m^2H^2 + \left(\frac{c_1(E)}{\operatorname{rk} E} - \frac{1}{2}K_X\right).Hm + \frac{\chi(E)}{\operatorname{rk} E}.$$

From these formulae one can read off the explicit inequalities for (H,A)-(semi)stability of sheaves on surfaces defined in the last chapter. The condition for (H,A)-semistability can be found already in [MW97].

Lemma 5.1.1 *Let E be a pure two-dimensional sheaf and $F \subset E$ a nontrivial subsheaf. Then $p_{H,A}(F) \leq_0 p_{H,A}(E)$ if and only if*

$$\left(\mu_H(F), \frac{\chi(F)}{\operatorname{rk} F}, -\mu_A(F)\right) \leq \left(\mu_H(E), \frac{\chi(E)}{\operatorname{rk} E}, -\mu_A(E)\right),$$

and $p_H(F) \leq p_H(E)$ if and only if $\left(\mu_H(F), \frac{\chi(F)}{\operatorname{rk} F}\right) \leq \left(\mu_H(E), \frac{\chi(E)}{\operatorname{rk} E}\right)$, where we consider the lexicographic ordering on $\mathbb{Q}^2$ and $\mathbb{Q}^3$.

Lemma 5.1.2 *Let $r \geq 2$ and $\Delta > 0$ be integers, so that we have the notion of a general ample divisor (see section 1.4.1). Let A and B be two general ample divisors such that there is a unique nongeneral ample divisor $H \in [A,B]$, and assume there is an A-stable and H-semistable sheaf E of rank r and discriminant Δ. Then E is (H,B)-stable.*

5.1. SEMISTABLE SHEAVES

Proof. Let $E' \subset E$ be a proper nontrivial subsheaf with $p_H(E') = p_H(E)$. In particular, $\mu_H(E') = \mu_H(E)$, hence the linear map

$$[A, B] \to \mathbb{R}, \quad h \mapsto \mu_h(E') - \mu_h(E)$$

is either zero everywhere or changes the sign when passing through H. In the first case one has $\mu_A(E') = \mu_A(E)$ and therefore $\frac{\chi(E')}{\operatorname{rk} E'} < \frac{\chi(E)}{\operatorname{rk} E}$ by the A-stability of E. In the second case one has $\mu_A(E') < \mu_A(E)$ and therefore $\mu_B(E') > \mu_B(E)$. By the characterisation in lemma 5.1.1 one has that $p_{H,B}(E') <_0 p_{H,B}(E)$ in both cases. $\square$

Note that one can always choose such A and B given any nongeneral ample divisor H if one replaces H by a sufficiently high multiple of H, which does not affect the stability notion. We will tacitly assume this replacement whenever necessary.

The proof of lemma 1.4.9 yields already its generalisation to (H, A)-stable sheaves:

Lemma 5.1.3 *Let $r \geq 2$ and $\Delta > 0$ be integers, H and A two ample divisors and B a general ample divisor, and assume there is an (H, A)-stable and μ_B-semistable sheaf E of rank r and discriminant Δ. Then E is B-stable.*

We restrict to K3 surfaces. The following proposition gives the key fact why to prefer a general ample divisor. The proportionality of the Mukai vector for destabilising subsheaves is essential in the local analysis of the singularities of moduli spaces of H-semistable sheaves, and in order to generalise the proof to (H, A)-semistable sheaves we need to establish this property.

Proposition 5.1.4 *Let X be a K3 surface, F a coherent sheaf on X of positive rank and H and A two ample divisors on X.*

1. *If A is $v(F)$-general, F (H, A)-semistable and $F' \subset F$ a nontrivial proper*

subsheaf with $p_{H,A}(F') = p_{H,A}(F)$ then

$$\frac{v(F')}{\text{rk } F'} = \frac{v(F)}{\text{rk } F}.$$

2. If A is $v(F)$-general and F is (H, A)-semistable with $v(F)$ primitive then F is (H, A)-stable.

3. Assume that H is $v(F)$-general. Then F is (H, A)-(semi)stable if and only if it is H-(semi)stable.

Proof.

1. By lemma 4.2.12 F' is saturated, so $0 < \text{rk } F' < \text{rk } F$. By lemma 5.1.1 one has
$$\left(\mu_H(F'), \frac{\chi(F')}{\text{rk } F'}, -\mu_A(F')\right) = \left(\mu_H(F), \frac{\chi(F)}{\text{rk } F}, -\mu_A(F)\right),$$
and as F is μ_H-semistable, we can apply corollary 1.4.4, i.e.
$$\frac{c_1(F')}{\text{rk } F'} = \frac{c_1(F)}{\text{rk } F}.$$
The claim follows using $v(F) = (\text{rk } F, c_1(F), \chi(F) - \text{rk } F)$.

2. As $v(F)$ is primitive clearly $(\text{rk } F, c_1(F), \chi(F))$ is primitive as well. Assume $E \subset F$ is a nontrivial saturated proper subsheaf with $p_{H,A}(E) = p_{H,A}(F)$, so
$$\text{rk } E(c_1(F), \chi(F)) = \text{rk } F(c_1(E), \chi(E))$$
by item 1. Thus
$$\frac{\text{rk } E}{\gcd(\text{rk } E, \text{rk } F)}(c_1(F), \chi(F)) = \frac{\text{rk } F}{\gcd(\text{rk } E, \text{rk } F)}(c_1(E), \chi(E)).$$
Clearly
$$1 \leq \gcd(\text{rk } E, \text{rk } F) \leq \text{rk } E < \text{rk } F,$$

5.1. SEMISTABLE SHEAVES

hence
$$\frac{\operatorname{rk} F}{\gcd(\operatorname{rk} E, \operatorname{rk} F)} > 1$$
and this integer divides $(c_1(F), \chi(F))$ and $\operatorname{rk} F$, which is a contradiction to $(\operatorname{rk} F, c_1(F), \chi(F))$ being primitive.

3. Let F be H-semistable. By definition one only has to show that

   a) F is already (H, A)-semistable and

   b) if F is (H, A)-stable then F is H-stable.

   Let $E \subset F$ be a nontrivial saturated proper subsheaf with $p_H(E) = p_H(F)$. Then recalling lemma 4.2.10 by item 1 one has $\mu_A(E) = \mu_A(F)$. Using lemma 5.1.1 this shows the first claim and gives a contradiction proving the second claim. □

The third part of this proposition sheds more light into the relation of H-semistability and (H, A)-semistability. In particular, (H, A)-semistability is only of interest for nongeneral H.

Proposition 5.1.5 *Let H and A be two ample divisors on X, $v \in \Lambda(X)$ primitive, $m \in \mathbb{N}$ and E an (H, A)-stable sheaf with $v(E) = mv$. Then $v^2 \geq -2$.*

Moreover, if $v^2 = -2$ then $m = 1$, and if in addition F is any (H, A)-semistable sheaf with $v(F) = v$ then $F \cong E$.

Proof. The proof carries over literally from proposition 1.3.7 using corollary 4.2.19 instead of [HL97] corollary 1.2.8, and proposition 4.2.18 instead of [HL97] proposition 1.2.7. □

Lemma 5.1.6 *Let X be a projective K3 surface, $v = (v_0, v_1, v_2) \in \Lambda(X)$ primitive with $v_0 > 0$, $m \in \mathbb{N}$, H an ample divisor and A an mv-general ample divisor. Let E be an (H, A)-semistable sheaf with Mukai vector $v(E) = mv$. Then $v^2 \geq -2$. In the case of equality E is Seshadri equivalent to a sheaf $F^{\oplus m}$ with F the unique (up to isomorphism) (H, A)-stable sheaf of Mukai vector*

$v(F) = v$. In particular, there is no (H, A)-stable sheaf with Mukai vector mv for $m \geq 2$ if $v^2 = -2$.

Proof. The proof carries over literally from proposition 1.5.1 using now proposition 5.1.4, corollary 4.2.19 and proposition 5.1.5. □

5.2 The moduli space $M_{H,A}(v)$

We come to the moduli space of (H, A)-semistable sheaves and state the analogous results to $M_H(v)$ whenever possible. Let X be a projective K3 surface, H and A two ample divisors on X and $v = (v_0, v_1, v_2) \in \Lambda(X)$ with $v_0 > 0$. The somehow traditional restriction to $v_0 > 0$ is just for simplicity as we don't need the case of $v_0 = 0$.

The Hilbert polynomial associated to the Mukai vector v, i.e. the Hilbert polynomial of any sheaf with Mukai vector v, is

$$P(m) := \int_X v \cdot \exp(mH) \cdot \sqrt{\text{td}(X)} = v_0 H^2 \frac{m^2}{2} + v_1.Hm + (v_2 + v_0).$$

As the map

$$c_1 : \text{Pic}(X) \to \text{NS}(X)$$

is a group isomorphism the sheaves with Mukai vector v are exactly those with Hilbert polynomial P and determinant $Q := c_1^{-1}(v_1)$, so we can define

$$M_{H,A}(v) := M(Q)$$

where $M(Q)$ is the fibre of the morphism $\det : M_{H,A}(P) \to \text{Pic}(X)$ over Q. This moduli space parametrises (H, A)-polystable sheaves with Mukai vector v. Let $M_{H,A}^s(v)$ denote the open subset of (H, A)-stable sheaves.

Recall the construction of the moduli space $M_{H,A}(P)$ in section 4.6. Let

$$\det : \text{Quot}(\mathcal{H}, P) \to \text{Pic}(X)$$

5.2. THE MODULI SPACE $M_{H,A}(V)$

be the determinant morphism and $\mathrm{Quot}(\mathcal{H}, P)_Q := \det^{-1}(Q)$ the fibre over a line bundle Q. Define

$$\begin{aligned} R(v) &:= R \cap \mathrm{Quot}(\mathcal{H}, P)_Q, \\ R^s(v) &:= R^s \cap \mathrm{Quot}(\mathcal{H}, P)_Q \text{ and} \\ R^{ss}(v) &:= R^{ss} \cap \mathrm{Quot}(\mathcal{H}, P)_Q. \end{aligned}$$

These schemes are $\mathrm{Gl}(V)$-invariant, and

$$M := M_{H,A}(v) = R^{ss}(v)/\!/\mathrm{PGl}(V) \quad \text{and} \quad M^s_{H,A}(v) = R^s(v)/\!/\mathrm{PGl}(V)$$

as by [HL97] section 4.3 it amounts to the same if one replaces $\mathrm{Sl}(V)$ by $\mathrm{PGl}(V)$.

By lemma 4.6.2 the stabiliser subgroup $\mathrm{PGl}(V)_{[q]}$ of a point $[q : \mathcal{H} \to E]$ with (H, A)-polystable E is isomorphic to $\mathrm{PAut}(E)$. Luna's Etale Slice Theorem [HL97] 4.2.12 yields a $\mathrm{PAut}(E)$-invariant locally closed subscheme $S \subset R^{ss}(v)$ through $[q]$ and an étale morphism

$$S/\!/\mathrm{PAut}(E) \to M.$$

For more details see [Leh02] section 3.3 or [KLS06] section 2.5. In particular, there is the Kuranishi map κ, a linear $\mathrm{PAut}(E)$-equivariant map

$$\kappa : \mathrm{Ext}^2(E, E)^*_0 \to \mathbb{C}[\mathrm{Ext}^1(E, E)]\hat{\,}$$

such that

$$\hat{\mathcal{O}}_{M,[E]} \cong \left(\mathbb{C}[\mathrm{Ext}^1(E, E)]\hat{\,}/(\mathrm{Im}\kappa) \right)^{\mathrm{PAut}(E)}.$$

Proposition 5.2.1 *$M^s_{H,A}(v)$ is nonsingular and each connected component has dimension $2 + v^2$.*

Proof. Any (H, A)-stable sheaf E is simple by corollary 4.2.19, thus

$$\mathrm{ext}^2(E, E)_0 = \mathrm{hom}(E, E)_0 = \mathrm{hom}(E, E) - 1 = 0,$$

hence by theorem 4.8.3 $M_{H,A}(v)$ is nonsingular in $[E]$ and

$$\dim_{[E]} M_{H,A}(v) = \mathrm{ext}^1(E,E)_0 = \mathrm{ext}^1(E,E).$$

The claim follows from $v^2 = -\chi(E,E)$. □

Theorem 5.2.2 $M^s_{H,A}(v)$ *admits a symplectic form.*

Proof. The construction of the symplectic form on $M^s_{H,A}(v)$ carries over from [HL97] chapter 10. □

Theorem 5.2.3 *If $M \subseteq M_{H,A}(v)$ is a connected component with $M \subseteq M^s_{H,A}(v)$ then one already has $M = M_{H,A}(v)$.*

Proof. The proof of [KLS06] theorem 4.1 carries over literally. □

Corollary 5.2.4 *If $M^s_{H,A}(v) = M_{H,A}(v)$ then $M_{H,A}(v)$ is irreducible.*

Proof. Any connected component of $M_{H,A}(v)$ is contained in $M^s_{H,A}(v)$, hence $M_{H,A}(v)$ is connected. Furthermore, $M^s_{H,A}(v)$ is nonsingular, hence $M_{H,A}(v)$ is irreducible. □

Theorem 5.2.5 *Let X be a projective K3 surface, $v = (v_0, v_1, v_2) \in \Lambda(X)$ primitive with $v_0 > 0$, $m \in \mathbb{N}$, H an ample divisor on X and A an mv-general ample divisor on X. Assume that $M^s_{H,A}(mv)$ is nonempty. Then $v^2 \geq -2$.*

1. *If $v^2 = -2$ then $m = 1$ and $M_{H,A}(v)$ consists of a reduced point.*

2. *If $v^2 = 0$ then $M_{H,A}(mv) = M^s_{H,A}(mv)$, and $M_{H,A}(mv)$ is a projective symplectic nonsingular surface.*

3. *Let $v^2 \geq 2$ and $M^s_{H,A}(v)$ be nonempty. Then $M_{H,A}(mv)$ is a projective symplectic variety of dimension $2 + m^2 v^2$.*

    a) *If $m = 1$ then $M_{H,A}(v) = M^s_{H,A}(v)$, and $M_{H,A}(v)$ is nonsingular.*

5.2. THE MODULI SPACE $M_{H,A}(V)$

b) If $m \geq 2$ then the singular locus of $M_{H,A}(mv)$ is nonempty and equals the strictly semistable locus.

   i. If $m = 2$ and $v^2 = 2$ then the singular locus has codimension 2 and $M_{H,A}(mv)$ admits a symplectic resolution.

   ii. If $m = 2$ and $v^2 > 2$ or $m > 2$ then $M_{H,A}(mv)$ is locally factorial, the singular locus has codimension at least 4 and the singularities are terminal. There is no open neighbourhood of a singular point that admits a symplectic resolution.

4. Let $v^2 \geq 2$ but now $M^s_{H,A}(v)$ be empty. Then $M_{H,A}(v)$ is empty as well, i.e. $m > 1$ by assumption. If $m = 2$ or 3 then $M_{H,A}(mv) = M^s_{H,A}(mv)$, and $M_{H,A}(mv)$ is a nonsingular projective symplectic variety of dimension $2 + m^2v^2$.

Proof. By lemma 5.1.6 one has $v^2 \geq -2$. Every connected component of $M^s_{H,A}(mv)$ is nonsingular of dimension $2 + m^2v^2$ by theorem 5.2.1 and carries a symplectic form by theorem 5.2.2. If $m = 1$ then $M^s_{H,A}(v) = M_{H,A}(v)$ by proposition 5.1.4. If $M^s_{H,A}(mv) = M_{H,A}(mv)$ then $M_{H,A}(mv)$ is irreducible by corollary 5.2.4. In particular, $M_{H,A}(v)$ is a projective symplectic nonsingular variety.

1. Let $v^2 = -2$. By lemma 5.1.6 one has $m = 1$ and all (H, A)-semistable sheaves with Mukai vector v are (H, A)-stable and isomorphic.

2. Let $v^2 = 0$. The proof of proposition 1.6.4 item 4a can be generalised to hold here as well.

3. Let $v^2 \geq 2$ and $M^s_{H,A}(v)$ be nonempty. The case of $m = 1$ is clear by the above statements, so let $m \geq 2$. The results herein are straightforward generalisations of [KLS06] and carry over literally as they are based on a local analysis using proposition 5.1.4 and the local description of $M_{H,A}(mv)$ is analogous as indicated above.

Theorem 5.2.6 $M_{H,A}(mv)$ *is a normal irreducible variety of dimension* $2 + v^2$.

Proof. See [KLS06] theorem 4.4. □

Theorem 5.2.7 *Assume* $m = 2$ *and* $v^2 > 2$ *or* $m > 2$. *Then* $M_{H,A}(mv)$ *is locally factorial.*

Proof. See [KLS06] theorem 5.3. □

Proposition 5.2.8 *The singular locus* $M_{H,A}(mv)_{sing}$ *of* $M_{H,A}(mv)$ *is non-empty and equals the strictly semistable locus. If* $m = 2$ *and* $v^2 = 2$ *then* codim $M_{H,A}(mv)_{sing} = 2$, *otherwise* ≥ 4.

Proof. See [KLS06] proposition 6.1. □

Theorem 5.2.9 *Assume* $m = 2$ *and* $v^2 > 2$ *or* $m > 2$. *Then* $M_{H,A}(mv)$ *is a locally factorial symplectic variety of dimension* $2+v^2$. *The singular locus is nonempty and has codimension at least 4. All singularities are symplectic but there is no open neighbourhood of a singular point that admits a symplectic resolution.*

Proof. See [KLS06] theorem 6.2. □

If $m = 2$ and $v^2 > 2$ or $m > 2$ the singularities are terminal by proposition 2.3.2.

We come to the case $m = 2$ and $v^2 = 2$.

Theorem 5.2.10 *Assume that* $m = 2$ *and* $v^2 = 2$. *Then the blow-up of* $M_{H,A}(mv)$ *along its reduced singular locus is a symplectic resolution.*

5.3. TERMINALISATIONS FOR $M_H(V)$

Proof. The proof carries over from [LS06] theorem 1.1. □

4. Let $v^2 \geq 2$ and $M^s_{H,A}(v)$ be empty. If $m = 2$ or 3 then $M_{H,A}(mv) = M^s_{H,A}(mv)$, and the claim follows from the statements at the beginning of the proof.

This finishes the proof of theorem 5.2.5. □

5.3 Terminalisations for $M_H(v)$

The moduli spaces $M_{H,A}(v)$ and their symplectic resolutions are good candidates for $\mathbb{Q}$-factorial projective symplectic terminalisations of $M_H(v)$ due to theorem 5.2.5 and the following result.

Proposition 5.3.1 *Let X be a projective K3 surface, $v = (v_0, v_1, v_2) \in \Lambda(X)$ with $v_0 > 0$, H and A two ample divisors. Assume that $M^s_H(v)$ is nonempty. Then there is a projective birational morphism*

$$f : M \to \overline{M^s_H(v)},$$

where M is the closure of the open subset of $M^s_{H,A}(v)$ of H-stable sheaves in $M_{H,A}(v)$. If $M_{H,A}(v)$ is irreducible then $M = M_{H,A}(v)$, and $\overline{M^s_H(v)}$ is irreducible as well.
Moreover, if $v_0 = 1$ then $f : M_{H,A}(v) \to M_B(v)$ is an isomorphism for any choice of an ample divisor B.

Proof. Proposition 3.1.6 yields the claimed projective birational morphism using observation 4.2.9 and the claimed isomorphism for $v_0 = 1$ using corollary 4.2.17. □

Hence we can reduce our question on $M_H(v)$ to the investigation of $M_{H,A}(v)$:

68 CHAPTER 5. MODULI SPACES FOR TWO-DIMENSIONAL SHEAVES

Proposition 5.3.2 *Let X be a projective K3 surface, $v = (v_0, v_1, v_2) \in \Lambda(X)$ with $v_0 \geq 2$, H a not v-general and A a v-general ample divisor. Assume that $M_H^s(v)$ is nonempty and that there is a $\mathbb{Q}$-factorial projective symplectic terminalisation*

$$\tilde{M}_{H,A}(v) \to M_{H,A}(v).$$

Then there is a $\mathbb{Q}$-factorial projective symplectic terminalisation

$$f : \tilde{M}_{H,A}(v) \to \overline{M_H^s(v)}.$$

1. *If there is a projective symplectic resolution $\tilde{M} \to \overline{M_H^s(v)}$ and $\tilde{M}$ can be chosen to be an irreducible symplectic manifold then $\tilde{M}$ is unique up to deformation.*

2. *If $\tilde{M}_{H,A}(v)$ is singular then $\overline{M_H^s(v)}$ admits no projective symplectic resolution.*

Proof. Concatenating the terminalisation of $M_{H,A}(v)$ with the morphism of proposition 5.3.1 yields the $\mathbb{Q}$-factorial projective symplectic terminalisation of $\overline{M_H^s(v)}$. Thus item 1 follows from theorem 1.1.4 and item 2 from proposition 2.3.4 □

Corollary 5.3.3 *Let X be a projective K3 surface, $v = (v_0, v_1, v_2) \in \Lambda(X)$ with $v_0 \geq 2$, H a not v-general ample divisor on X and M an irreducible component of $M_H(v)$ containing no H-stable sheaves.*

Assume that for all $w = (w_0, w_1, w_2) \in \Lambda(X)$ with $1 < w_0 < v_0$ and such that H is not w-general, $\frac{w_1.H}{w_0} = \frac{v_1.H}{v_0}$ and $\frac{w_2}{w_0} = \frac{v_2}{v_0}$ there is a $\mathbb{Q}$-factorial projective symplectic terminalisation of $M_{H,A_w}(w)$ for a suitable w-general ample divisor A_w. Then there is a $\mathbb{Q}$-factorial projective symplectic terminalisation $\tilde{M} \to M$.

If $\tilde{M}$ can be chosen to be an irreducible symplectic manifold then it is deformation equivalent to some symplectic resolution of some $M_{H,A}(w)$, where $w = (w_0, w_1, w_2) \in \Lambda(X)$ has the above properties and A is a w-general ample divisor, to a symplectic resolution of some $M_H(w)$, where $1 \leq w_0 < v_0$, H is

5.3. TERMINALISATIONS FOR $M_H(V)$

w-general, $\frac{w_1.H}{w_0} = \frac{v_1.H}{v_0}$ and $\frac{w_2}{w_0} = \frac{v_2}{v_0}$, or to a Hilbert scheme of points on a K3 surface.

Proof. Consider the decomposition $v = \sum_{i=1}^{m} n_i v^{(i)}$ given by proposition 2.1.1. As the Mukai vectors $v^{(i)}$ for $1 \leq i \leq m$ belong to H-stable direct summands of a strictly H-polystable sheaf with Mukai vector v one has $1 < v_0^{(i)} < v_0$,

$$\frac{v_1^{(i)}.H}{v_0^{(i)}} = \frac{v_1.H}{v_0} \quad \text{and} \quad \frac{v_2^{(i)}}{v_0^{(i)}} = \frac{v_2}{v_0}$$

for all $1 \leq i \leq m$. If H is $v^{(i)}$-general for some $1 \leq i \leq m$ then $M_H(v^{(i)})$ is a symplectic variety that admits a $\mathbb{Q}$-factorial projective symplectic terminalisation by proposition 1.6.4. Note that for rank one every ample divisor is general. Thus we have established the assumption of theorem 2.3.5 item 1, which hence yields the claim. $\square$

Unfortunately, theorem 5.2.5 does not give a complete answer and the assumptions are not necessarily satisfied.

Question 5.3.4 *Let X be a projective K3 surface, $v = (v_0, v_1, v_2) \in \Lambda(X)$ primitive with $v_0 > 0$ and $v^2 \geq 2$, $4 \leq m \in \mathbb{N}$, H a not mv-general ample divisor on X and A an mv-general ample divisor on X. Assume that $M_{H,A}^s(mv)$ is nonempty.*

1. *Is $M_{H,A}^s(v)$ nonempty?*

2. *If $M_{H,A}^s(v)$ is empty, is there a $\mathbb{Q}$-factorial projective symplectic terminalisation of $M_{H,A}(mv)$?*

If one could close this gap then any projective irreducible symplectic manifold lying birationally over any component of some $M_H(v)$ always arises also as a symplectic resolution of some $M_{H,A}(v)$. This leads us to the next question.

Question 5.3.5 *Do the moduli spaces $M_{H,A}(v)$ yield new examples of projective irreducible symplectic manifolds?*

Due to theorem 1.1.4 one can reduce this question to the existence of birational maps.

5.4 Birational maps between moduli spaces

It is not possible in general to construct morphisms between moduli spaces of semistable sheaves of positive rank for different choices of ample divisors, so one has to settle for birational maps. Fortunately one knows that birational projective irreducible symplectic manifolds are deformation equivalent, as stated in theorem 1.1.4.

Let X be a projective K3 surface and $v = (v_0, v_1, v_2) \in \Lambda(X)$ with $v_0 > 0$.

Proposition 5.4.1 *Let (H_i, A_i) be a pair of two ample divisors for $i = 1, 2$. Then the two open subsets $U_1 \subset M^s_{H_1, A_1}(v)$ and $U_2 \subset M^s_{H_2, A_2}(v)$ containing all sheaves that are (H_i, A_i)-stable for $i = 1, 2$ are isomorphic.*

Proof. Let $\pi : R^s(v) \to M^s_{H_1, A_1}(v)$ be the quotient map, see the construction of the moduli space in section 5.2. $R^s(v)$ carries a universal family. By proposition 4.4.2 the maximal subset $W \subset R^s(v)$ parametrising only (H_2, A_2)-stable sheaves is open. As W is $\mathrm{PGl}(V)$-invariant $U_1 = \pi(W)$ is the open subset of $M^s_{H_1, A_1}(v)$ containing all sheaves that are (H_i, A_i)-stable.

The universal family on R^s restricted to W induces a morphism $f : W \to M^s_{H_2, A_2}(v)$ by the universal property of coarse moduli spaces, and its image is U_2. This morphism is $\mathrm{PGl}(V)$-invariant and hence descends to a morphism $\overline{f} : U_1 \to U_2$. Exchanging 1 and 2 yields the inverse morphism. □

Corollary 5.4.2 *Let H be an ample divisor on X and A and B two v-general ample divisors on X. Assume there is a coherent sheaf F with Mukai vector v that is (H, A)-stable and B-stable.*

1. *If $M_{H,A}(v)$ is irreducible then $M_{H,A}(v)$ and $M_B(v)$ are birational.*

2. *Let v be primitive.*

a) If $v^2 = 0$ then $M_{H,A}(v)$ is a projective K3 surface.

b) If $v^2 \geq 2$ then $M_{H,A}(v)$ is deformation equivalent to $\text{Hilb}^{\frac{v^2}{2}+1}(X)$.

Proof.

1. By proposition 1.6.4 $M_B(v)$ is also irreducible, so nonempty open subsets are dense and the isomorphism of proposition 5.4.1 gives the desired birational map.

2. By theorem 5.2.5 $M_{H,A}(v)$ is irreducible, so one has that $M_{H,A}(v)$ is birational to $M_B(v)$ in both cases as just seen. The claim follows from proposition 1.6.4 together with theorem 1.1.4. For the first case note that an irreducible symplectic surface is already K3. □

5.5 On the discriminant

We need some preliminaries before we can deduce an existence result for the stable sheaves needed in corollary 5.4.2.

Let X be a nonsingular projective surface with an ample divisor H and F a coherent sheaf on X. Recall that

$$\Delta(F) := 2\text{rk}\, F c_2(F) - (\text{rk}\, F - 1)c_1(F)^2 = c_1(F)^2 - 2\text{rk}\, F \text{ch}_2(F) \quad (5.1)$$

is called the discriminant of F, see e.g. [HL97] section 3.4, and if X is a K3 surface then

$$\Delta(F) = v(F)^2 + 2(\text{rk}\, F)^2. \quad (5.2)$$

Warning: *There are different conventions of the definition of the discriminant in the literature!*

The following statement is contained in [HL97] corollary 7.3.2. As the calculation is omitted therein, we give the details here.

Lemma 5.5.1 *Let $0 = F_0 \subset F_1 \subset ... \subset F_n = F$ be a filtration of a coherent sheaf F on X with positive rank such that the graded objects $gr_i := F_i/F_{i-1}$ have positive rank for $i = 1, ..., n$. Then*

$$\sum_{i=1}^n \frac{\Delta(gr_i)}{\operatorname{rk} gr_i} - \frac{\Delta(F)}{\operatorname{rk} F} = \sum_{i<j} \frac{\operatorname{rk} gr_i \operatorname{rk} gr_j}{\operatorname{rk} F} \left(\frac{c_1(gr_i)}{\operatorname{rk} gr_i} - \frac{c_1(gr_j)}{\operatorname{rk} gr_j} \right)^2.$$

Proof. We calculate

$$\frac{1}{2\operatorname{rk} F}(c_1(F)^2 - \Delta(F)) \stackrel{(5.1)}{=} \operatorname{ch}_2(F) = \sum_{i=1}^n \operatorname{ch}_2(gr_i)$$

$$\stackrel{(5.1)}{=} \sum_{i=1}^n \frac{1}{2\operatorname{rk} gr_i}(c_1(gr_i)^2 - \Delta(gr_i))$$

and

$$\sum_{i=1}^n \frac{\Delta(gr_i)}{\operatorname{rk} gr_i} - \frac{\Delta(F)}{\operatorname{rk} F}$$

$$= \sum_{i=1}^n \frac{c_1(gr_i)^2}{\operatorname{rk} gr_i} - \frac{c_1(F)^2}{\operatorname{rk} F}$$

$$= \sum_{i,j=1}^n \left(\frac{\operatorname{rk} gr_j}{\operatorname{rk} F} \frac{c_1(gr_i)^2}{\operatorname{rk} gr_i} - \frac{c_1(gr_i) c_1(gr_j)}{\operatorname{rk} F} \right)$$

$$= \sum_{i<j} \left(\frac{\operatorname{rk} gr_j}{\operatorname{rk} F} \frac{c_1(gr_i)^2}{\operatorname{rk} gr_i} + \frac{\operatorname{rk} gr_i}{\operatorname{rk} F} \frac{c_1(gr_j)^2}{\operatorname{rk} gr_j} - 2\frac{c_1(gr_i) c_1(gr_j)}{\operatorname{rk} F} \right)$$

$$= \sum_{i<j} \frac{\operatorname{rk} gr_i \operatorname{rk} gr_j}{\operatorname{rk} F} \left(\frac{c_1(gr_i)^2}{(\operatorname{rk} gr_i)^2} + \frac{c_1(gr_j)^2}{(\operatorname{rk} gr_j)^2} - 2\frac{c_1(gr_i) c_1(gr_j)}{\operatorname{rk} gr_i \operatorname{rk} gr_j} \right)$$

$$= \sum_{i<j} \frac{\operatorname{rk} gr_i \operatorname{rk} gr_j}{\operatorname{rk} F} \left(\frac{c_1(gr_i)}{\operatorname{rk} gr_i} - \frac{c_1(gr_j)}{\operatorname{rk} gr_j} \right)^2.$$

□

5.5. ON THE DISCRIMINANT

Corollary 5.5.2 *If all gr_i have the same slope with respect to H then one has*

$$\sum_{i=1}^{n} \frac{\Delta(gr_i)}{\operatorname{rk} gr_i} \leq \frac{\Delta(F)}{\operatorname{rk} F}.$$

Moreover, if X is a K3 surface and $\frac{c_1(gr_i)}{\operatorname{rk} gr_i} \neq \frac{c_1(gr_j)}{\operatorname{rk} gr_j}$ for all $1 \leq i < j \leq n$ then one even has

$$\sum_{i=1}^{n} \frac{\Delta(gr_i)}{\operatorname{rk} gr_i} \leq \frac{\Delta(F)}{\operatorname{rk} F} - \sum_{i<j} \frac{2 \operatorname{rk} gr_i \operatorname{rk} gr_j}{\operatorname{rk} F \operatorname{lcm}(\operatorname{rk} gr_i, \operatorname{rk} gr_j)^2}$$

$$\leq \frac{\Delta(F)}{\operatorname{rk} F} - \sum_{i<j} \frac{2}{\operatorname{rk} F \operatorname{rk} gr_i \operatorname{rk} gr_j},$$

where lcm *denotes the* **l**east **c**ommon **m**ultiple.

Proof. By assumption one has

$$\left(\frac{c_1(gr_i)}{\operatorname{rk} gr_i} - \frac{c_1(gr_j)}{\operatorname{rk} gr_j} \right).H = 0,$$

hence

$$\left(\frac{c_1(gr_i)}{\operatorname{rk} gr_i} - \frac{c_1(gr_j)}{\operatorname{rk} gr_j} \right)^2 \leq 0$$

for all i, j by the Hodge index theorem, see e.g. [HL97] theorem V.1.9. If X is a K3 surface then the intersection pairing is even and nondegenerate, and therefore even

$$\left(\frac{c_1(gr_i)}{\operatorname{rk} gr_i} - \frac{c_1(gr_j)}{\operatorname{rk} gr_j} \right)^2$$

$$= \frac{1}{\operatorname{lcm}(\operatorname{rk} gr_i, \operatorname{rk} gr_j)^2} \left(\operatorname{lcm}(\operatorname{rk} gr_i, \operatorname{rk} gr_j) \left(\frac{c_1(gr_i)}{\operatorname{rk} gr_i} - \frac{c_1(gr_j)}{\operatorname{rk} gr_j} \right) \right)^2$$

$$\leq -\frac{2}{\operatorname{lcm}(\operatorname{rk} gr_i, \operatorname{rk} gr_j)^2}.$$

□

Lemma 5.5.3 Let X be a $K3$ surface with ample divisor H, $2 \leq n \in \mathbb{N}$ and $0 = F_0 \subset F_1 \subset ... \subset F_n = F$ a filtration of a coherent sheaf F on X with positive rank r such that all $gr_i = F_i/F_{i-1}$ have positive rank r_i, are μ_H-semistable, have the same slope with respect to H and $\frac{c_1(gr_i)}{\operatorname{rk} gr_i} \neq \frac{c_1(gr_j)}{\operatorname{rk} gr_j}$ for all $1 \leq i < j \leq n$. Then

$$\sum_{i<j} \chi(gr_i, gr_j) \leq -\frac{\Delta(F)}{2r}(n-1) + r^2 - \sum_{i=1}^{n} r_i^2 - \frac{r-n+1}{r}\sum_{i<j}\frac{1}{r_i r_j}.$$

Proof. One has $r_i \leq r - n + 1$ for all $i = 1, .., n$. Furthermore, $\Delta(gr_i) \geq 0$ by the Bogomolov inequality (see e.g. [HL97] theorem 3.4.1). So we can calculate

$$\begin{aligned}
\sum_{i<j} \chi(gr_i, gr_j) &\stackrel{\text{prop. 1.3.3}}{=} -\sum_{i<j} \langle v(gr_i), v(gr_j) \rangle \\
&= \frac{1}{2}\left(-\sum_{i,j=1}^{n} \langle v(gr_i), v(gr_j) \rangle + \sum_{i=1}^{n} v(gr_i)^2\right) \\
&= \frac{1}{2}\left(\sum_{i=1}^{n} v(gr_i)^2 - v(F)^2\right) \\
&\stackrel{(5.2)}{=} \frac{1}{2}\left(\sum_{i=1}^{n}(\Delta(gr_i) - 2r_i^2) - \Delta(F) + 2r^2\right) \\
&\leq \sum_{i=1}^{n}\frac{r-n+1}{2r_i}\Delta(gr_i) - \sum_{i=1}^{n}r_i^2 - \frac{\Delta(F)}{2} + r^2 \\
&\stackrel{\text{cor. 5.5.2}}{\leq} \frac{r-n+1}{2}\left(\frac{\Delta(F)}{r} - \sum_{i<j}\frac{2}{rr_i r_j}\right) - \sum_{i=1}^{n}r_i^2 - \frac{\Delta(F)}{2} + r^2 \\
&= -\frac{\Delta(F)}{2r}(n-1) + r^2 - \sum_{i=1}^{n}r_i^2 - \frac{r-n+1}{r}\sum_{i<j}\frac{1}{r_i r_j}.
\end{aligned}$$

$\square$

5.6 Existence of stable sheaves

Some parts of the proof of the following proposition are based on an idea we learned from an unpublished note of Christoph Sorger.

Proposition 5.6.1 *Let X be a K3 surface, $v = (v_0, v_1, v_2) \in \Lambda(X)$ with $v_0 \geq 2$ and H, A and B three ample divisors on X. Assume that $M_{A,B}^s(v)$ is nonempty and contains no H-semistable sheaves, and let $R^s \to M_{A,B}^s(v)$ be the geometric quotient of the construction of the moduli space $M_{A,B}(v)$ in section 5.2 and $F \in \mathrm{Coh}(R^s \times X)$ the universal quotient family. Then there is an open dense subset $S \subset R^s$ and a subsheaf $F' \subset F|_S$ such that for all $s \in S$ one has*

1. *an exact sequence $0 \to F'_s \to F_s \to F''_s \to 0$ on the fibre over s with*

2. $p_H(F'_s) > p_H(F_s) > p_H(F''_s)$,

3. $\hom(F'_s, F''_s) = 0$ *and*

4. $1 - \chi(F'_s, F''_s) = \mathrm{ext}^2_-(F_s, F_s) \leq \mathrm{end}(F'_s) + \mathrm{end}(F''_s)$,

where we calculate $\mathrm{ext}^2_-(F_s, F_s)$ with respect to the filtration $F'_s \subset F_s$ (for a definition see [HL97] section 2.A), and if $v_0 = 2$ then additionally

1. $\mathrm{ext}^2_-(F_s, F_s) = 2$,

2. $-\chi(F'_s, F''_s) = \mathrm{ext}^1(F'_s, F''_s) = 1$ *and*

3. $0 \geq (c_1(F'_s) - c_1(F''_s))^2 = v^2 - 4 = v(F'_s)^2 + v(F''_s)^2 - 2$.

Proof. By corollary 4.2.19 one has

$$\mathrm{ext}^2(F_s, F_s) = \hom(F_s, F_s) = 1 \tag{5.3}$$

for all $s \in S$. By the same arguments as in the proof of [HL97] theorem 10.2.1 R^s is nonsingular and the Kodaira-Spencer map κ is given by the concatenation of the two maps

$$T_s R^s \longrightarrow T_{[F_s]} M_A^s(v) \xrightarrow{\cong} \mathrm{Ext}^1(F_s, F_s).$$

Furthermore, the first map is surjective, hence κ is surjective as well.

In the following every notion is understood to be with respect to the ample divisor H whenever not explicitly stated differently. By [HL97] theorem 2.3.2 there is a relative Harder-Narasimhan filtration $F_\bullet$ and an open dense subscheme $S \subset R^s$ such that the restriction of the filtration to a fibre over $s \in S$ is a Harder-Narasimhan filtration of F_s. As the open subset of R^s containing H-semistable sheaves is empty the filtration is nontrivial. We only take the first step $F' := F_{\ell-1}|_S \subset F_\ell|_S = F|_S$ of the filtration restricted to S, which gives us an exact sequence

$$0 \to F'_s \to F_s \to F''_s \to 0 \tag{5.4}$$

on the fibres over $s \in S$ with

$$p_H(F'_s) > p_H(F_s) > p_H(F''_s). \tag{5.5}$$

By the proof of [HL97] theorem 2.3.2 one has

$$\hom(F'_s, F''_s) = 0 \tag{5.6}$$

and

$$\pi : \text{Quot}_{S \times X/S}(F, P_-) \to S$$

is an isomorphism, where $P_- := P_H(F''_s)$ (this is independent of s). Let $s \in S$ be a closed point and x be the unique point with $s = \pi(x)$, which corresponds to the exact sequence 5.4. By [HL97] theorem 2.2.7 the kernel of the obstruction map $\mathfrak{o} : T_s S \to \text{Ext}^1(F'_s, F''_s)$ is

$$\ker \mathfrak{o} = \text{Im } T_x \pi = \dim T_s S,$$

hence $\mathfrak{o}$ is the zero map. As $\mathfrak{o}$ is given by

$$\mathfrak{o} : T_s S \xrightarrow{\kappa} \text{Ext}^1(F_s, F_s) \xrightarrow{c} \text{Ext}^1(F'_s, F''_s)$$

5.6. EXISTENCE OF STABLE SHEAVES

and κ is surjective as explained above one has $c = 0$ as well. For the short filtration $0 \subset F'_s \subset F_s$ there is a long exact sequence

$$\ldots \to \mathrm{Ext}^i_-(F_s, F_s) \to \mathrm{Ext}^i(F_s, F_s) \to \mathrm{Ext}^i(F'_s, F''_s) \to \mathrm{Ext}^{i+1}_-(F_s, F_s) \to \ldots,$$

which decomposes to the exact sequence

$$0 \to \mathrm{Ext}^1(F'_s, F''_s) \to \mathrm{Ext}^2_-(F_s, F_s) \to \mathrm{Ext}^2(F_s, F_s) \to \mathrm{Ext}^2(F'_s, F''_s) \to 0$$

as $c = 0$. Thus

$$\begin{aligned} 0 &= -\mathrm{ext}^1(F'_s, F''_s) + \mathrm{ext}^2_-(F_s, F_s) - \mathrm{ext}^2(F_s, F_s) + \mathrm{ext}^2(F'_s, F''_s) \\ &\stackrel{(5.3),\,(5.6)}{=} \chi(F'_s, F''_s) + \mathrm{ext}^2_-(F_s, F_s) - 1\,. \end{aligned}$$

By [HL97] theorem 2.A.4 there is a spectral sequence

$$\mathrm{Ext}^{p+q}_-(F_s, F_s) \Leftarrow E_1^{pq} = \begin{cases} 0 & p < 0 \\ \prod_i \mathrm{Ext}^{p+q}(gr_i F_s, gr_{i-p} F_s) & p \geq 0 \end{cases}.$$

Hence

$$\begin{aligned} \mathrm{ext}^2_-(F_s, F_s) &\leq \sum_{i \geq j} \mathrm{ext}^2(gr_i F_s, gr_j F_s) \\ &= \sum_{i \geq j} \hom(gr_j F_s, gr_i F_s) \\ &= \mathrm{end}(F'_s) + \mathrm{end}(F''_s) + \hom(F'_s, F''_s) \\ &\stackrel{(5.6)}{=} \mathrm{end}(F'_s) + \mathrm{end}(F''_s) \end{aligned} \qquad (5.7)$$

Assume $v_0 = 2$. Then F'_s and F''_s are line bundles and therefore (A, B)-stable by corollary 4.2.17, hence simple by corollary 4.2.19. As F_s is (A, B)-stable, one has

$$p_{A,B}(F'_s) <_0 p_{A,B}(F_s) <_0 p_{A,B}(F''_s) \qquad (5.8)$$

and in particular,
$$0 = \hom(F_s'', F_s') = \ext^2(F_s', F_s'')$$
by proposition 4.2.18. The sequence
$$0 \to F_s' \to F_s \to F_s'' \to 0$$
gives a nontrivial element in $\Ext^1(F_s', F_s'')$, thus $\ext^1(F_s', F_s'') \geq 1$. Altogether one has
$$1 \leq \ext^1(F_s', F_s'') = -\chi(F_s', F_s'') = \ext^2_-(F_s, F_s) - 1 \overset{(5.7)}{\leq} 1,$$
which gives us equality everywhere. In particular,
$$\begin{aligned}
-2 &= 2\chi(F_s', F_s'') \\
&= -2\langle v(F_s'), v(F_s'') \rangle \\
&= v(F_s')^2 + v(F_s'')^2 - v^2 \\
&\overset{(5.2)}{=} \Delta(F_s') + \Delta(F_s'') - 4 - v^2 \\
&\overset{1.\ 5.5.1}{=} \frac{\Delta(F_s)}{2} + \frac{1}{2}\left(c_1(F_s') - c_1(F_s'')\right)^2 - 4 - v^2 \\
&\overset{(5.2)}{=} \frac{1}{2}\left(c_1(F_s') - c_1(F_s'')\right)^2 - \frac{v^2}{2}.
\end{aligned}$$
By the inequalities 5.5 and 5.8 one has
$$\begin{aligned}
\mu_H(F_s') &\geq \mu_H(F_s) \quad \text{and} \\
\mu_A(F_s') &\leq \mu_A(F_s),
\end{aligned}$$
hence there is an ample $\mathbb{Q}$-divisor $H' \in [H, A]$ such that
$$\mu_{H'}(F_s') = \mu_{H'}(F_s) = \mu_{H'}(F_s''),$$

5.6. EXISTENCE OF STABLE SHEAVES

and the Hodge index theorem (see e.g. [HL97] theorem V.1.9.) yields

$$0 \geq \bigl(c_1(F'_s) - c_1(F''_s)\bigr)^2 = v^2 - 4\,.$$

□

Theorem 5.6.2 *Let X be a K3 surface, $v = (v_0, v_1, v_2) \in \Lambda(X)$ with $v_0 \geq 2$ and*

$$v^2 > 2\left(v_0^3 - v_0^2 - v_0 - (v_0 - 1)\left\lfloor\frac{v_0^2}{4}\right\rfloor^{-1}\right) =: \varphi(v_0)\,,$$

H a not v-general ample divisor and A a v-general ample divisor in a chamber touching H. Then there is an A-stable and H-semistable sheaf with Mukai vector v.

Proof. One can easily verify that by the assumptions on v one has $v^2 \geq 2$, and therefore $M_A^s(v) \neq \emptyset$ by proposition 1.6.4. Set $B = A$ and assume that $M_A^s(v)$ contains no H-semistable sheaves. Then by proposition 5.6.1 there is an $[F] \in M_A^s(v)$ with an exact sequence

$$0 \to F' \to F \to F'' \to 0$$

such that

$$p_H(F') > p_H(F) > p_H(F'')\,, \tag{5.9}$$
$$\hom(F', F'') = 0 \quad \text{and} \tag{5.10}$$
$$1 - \chi(F', F'') = \mathrm{ext}_{-}^2(F, F) \leq \mathrm{end}(F') + \mathrm{end}(F'')\,. \tag{5.11}$$

80 CHAPTER 5. MODULI SPACES FOR TWO-DIMENSIONAL SHEAVES

As F is in particular μ_A-semistable, it is also μ_H-semistable by proposition 1.4.7. Together with inequality 5.9 one has

$$\mu_H(F') = \mu_H(F) = \mu_H(F'') \quad \text{and} \tag{5.12}$$

$$\frac{\chi(F')}{\operatorname{rk} F'} > \frac{\chi(F)}{\operatorname{rk} F} > \frac{\chi(F'')}{\operatorname{rk} F''}, \tag{5.13}$$

and F' and F'' are μ_H-semistable. Thus by [O'G96] lemma 1.7 one has

$$\operatorname{end}(F') \leq (\operatorname{rk} F')^2 \quad \text{and} \quad \operatorname{end}(F'') \leq (\operatorname{rk} F'')^2. \tag{5.14}$$

Moreover, the A-stability of F ensures

$$p_A(F') < p_A(F) < p_A(F''),$$

and because of inequality 5.13 even

$$\mu_A(F') < \mu_A(F) < \mu_A(F''),$$

and therefore

$$\frac{c_1(F')}{\operatorname{rk} F'} \neq \frac{c_1(F'')}{\operatorname{rk} F''}. \tag{5.15}$$

Altogether one has

$$
\begin{aligned}
1 &\stackrel{(5.11)}{\leq} \chi(F', F'') + \operatorname{end}(F') + \operatorname{end}(F'') \\
&\stackrel{\text{l. 5.5.3, (5.14)}}{\leq} -\frac{\Delta(F)}{2v_0} + v_0^2 - \frac{v_0 - 1}{v_0 \operatorname{rk} F' \operatorname{rk} F''} \\
&\stackrel{(5.2)}{\leq} -\frac{v^2}{2v_0} + v_0^2 - v_0 - \frac{v_0 - 1}{v_0 \operatorname{rk} F' \operatorname{rk} F''} \\
&\leq -\frac{v^2}{2v_0} + v_0^2 - v_0 - \frac{v_0 - 1}{v_0}\left[\frac{v_0^2}{4}\right]^{-1},
\end{aligned}
$$

5.6. EXISTENCE OF STABLE SHEAVES

where the last inequality follows from $\text{rk } F' \text{ rk } F'' \leq \left\lfloor \frac{v_0^2}{4} \right\rfloor$. Thus

$$v^2 \leq 2\left(v_0^3 - v_0^2 - v_0 - (v_0 - 1)\left\lfloor \frac{v_0^2}{4} \right\rfloor^{-1}\right)$$

in contradiction to the assumption of the theorem. □

Let us evaluate φ for small values:

n	2	3	4	5	6
$\varphi(n)$	2	28	86.5	188.7	346.9

Table 5.1: evaluation of φ

In particular, the only interesting exceptional case for rank two might occur for $v^2 = 2$. To realise this case one needs a K3 surface that holds a divisor D with $D^2 = -2$, $D.H = 0$, $D.A < 0$ and such that $D + v_1$ is divisible by 2 in $\text{NS}(X)$.

Lemma 5.6.3 *Let X be a K3 surface, S a scheme of finite type over $\mathbb{C}$ and $F \in \text{Coh}(S \times X)$ an S-flat family on X. Then the Mukai vector $v(F_s)$ is locally constant as a function of $s \in S$.*

Proof. By [HL97] proposition 2.1.2 the Hilbert polynomials $P_H(F_s)$ with respect to any fixed ample divisor H are locally constant as a function of $s \in S$. Thus $\text{rk } F_s$ and $\chi(F_s)$ are locally constant, and for all ample divisors H one has that $c_1(F_s).H$ is locally constant. As X is a K3 surface, the intersection pairing is nondegenerate, hence it is enough to show that $c_1(F_s).D$ is locally constant for any divisor D. But the ample cone is open in $\text{NS}(X)_\mathbb{Q}$ and $\text{NS}(X)$ is free, hence D is a linear combination of ample divisors. □

Proposition 5.6.4 *Let X be a K3 surface, $v \in \Lambda(X)$ and H, A and B three ample divisors such that $M_{A,B}^s(v)$ is connected and contains no H-semistable* sheaves. Then there is an open subset $U \subset M^s_{A,B}(v)$ and a morphism

$$f : U \to \prod_{i=1}^{\ell} M_H(v^{(i)})$$

for a suitable decomposition $v = \sum_{i=1}^{\ell} v^{(i)}$ that is induced by the Harder-Narasimhan filtration with respect to H.

Let $\ell = 2$ and $V \subset f^{-1}(\prod_{i=1}^{2} M^s_H(v^{(i)}))$ be the subset of all $[E]$ with Harder-Narasimhan filtration $0 = E_0 \subset E_1 \subset E_2 = E$ with respect to H such that one has $\mathrm{ext}^1(E_1, E/E_1) = 1$. Then $f|_V$ is injective. If furthermore V contains an open subset and $v^2 = \sum_{i=1}^{2} (v^{(i)})^2 + 2$ then

$$\dim f(V) = \dim \prod_{i=1}^{2} M_H(v^{(i)}).$$

In particular, $\overline{f(V)}$ is an irreducible component of $\prod_{i=1}^{\ell} M_H(v^{(i)})$.

Proof. Let $\pi : R^s \to M^s_{A,B}(v)$ be the geometric quotient of the construction of the moduli space $M_{A,B}(v)$ in section 5.2 and $F \in \mathrm{Coh}(R^s \times X)$ the universal quotient family of (A, B)-stable sheaves on X. By [HL97] theorem 2.3.2 there is a relative Harder-Narasimhan filtration $F_0 \subset \ldots \subset F_\ell$ of F and an open dense subscheme $S \subset R^s$ such that the restriction of the filtration to a fibre over $s \in S$ is a Harder-Narasimhan filtration of F_s with respect to H, and the restriction of the factors $gr_i := F_i/F_{i-1}$ to S are flat over S for all $1 \leq i \leq \ell$. S is connected because it is an open subset of R^s, which in turn is connected as it is a pricipal bundle over the connected variety $M^s_A(v)$ with connected structure group, hence by lemma 5.6.3 the Mukai vectors $v^{(i)} := v((gr_i)_s)$ are independent of $s \in S$ for all $1 \leq i \leq \ell$. S is an invariant open set, hence its image $U := \pi(S)$ is open. The families $gr_i|_S$ yield an invariant morphism

$$S \to \prod_{i=1}^{\ell} M_H(v^{(i)}),$$

5.6. EXISTENCE OF STABLE SHEAVES

which descends to a morphism

$$f : U \to \prod_{i=1}^{\ell} M_H(v^{(i)}).$$

Let $\ell = 2$, $[E], [E'] \in V$ with $f([E]) = f([E'])$ and $E_\bullet$ and $E'_\bullet$ their Harder-Narasimhan filtrations with respect to H. Then one has $E_1 \cong E_1/E_0 \cong E'_1/E'_0 \cong E'_1$ and two exact sequences

$$0 \to E_1 \to E \to E/E_1 \to 0 \quad \text{and}$$
$$0 \to E_1 \to E' \to E/E_1 \to 0.$$

As E is (H, A)-semistable one has $p_{A,B}(E) <_0 p_{A,B}(E/E_1)$ by corollary 4.2.16, and therefore $E \not\cong E_1 \oplus E/E_1$. Hence the first sequence is nonsplit and corresponds to a nontrivial element in $\text{Ext}^1(E/E_1, E_1)$. Analogously the second sequence corresponds to a nontrivial element. As

$$\text{ext}^1(E/E_1, E_1) = \text{ext}^1(E_1, E/E_1) = 1$$

one has $E \cong E'$. If V contains an open subset then $\dim f(V) = \dim V = v^2 + 2$. If furthermore $v^2 = \sum_{i=1}^{2}(v^{(i)})^2 + 2$ then

$$\dim f(V) = \sum_{i=1}^{2}(v^{(i)})^2 + 4 = \dim \prod_{i=1}^{2} M_H(v^{(i)}).$$

$M^s_{A,B}(v)$ is nonsingular by proposition 5.2.1 and connected by assumption, hence it is irreducible. Thus V, $f(V)$ and $\overline{f(V)}$ are irreducible, and in particular, $\overline{f(V)}$ is an irreducible component of $\prod_{i=1}^{\ell} M_H(v^{(i)})$. □

Corollary 5.6.5 *Let X be a K3 surface, $v = (2, v_1, \frac{v_1^2 - 2}{4}) \in \Lambda(X)$ and H, A and B three ample divisors such that $M^s_{A,B}(v)$ is nonempty, connected and contains no H-semistable sheaves. Then $M^s_{A,B}(v)$ is birational to $M_H(v') \times M_H(v'')$ with $v' = (1, v'_1, v'_2)$, $v'' = (1, v''_1, v''_2)$ and $v'^2 + v''^2 = 0$.*

Proof. This follows from proposition 5.6.4 using proposition 5.6.1. Note that $M_H(v') \times M_H(v'')$ is irreducible as the factors are irreducible by proposition 1.6.4. □

5.7 More results on $M_{H,A}(v)$

We now can give partial answers to our questions 5.3.4 and 5.3.5.

Theorem 5.7.1 *Let X be a projective K3 surface, $v = (v_0, v_1, v_2) \in \Lambda(X)$ primitive and $m \in \mathbb{N}$ with $mv_0 \geq 2$, H a not mv-general ample divisor on X and A an mv-general ample divisor on X in a chamber touching H.*

1. *If $(mv)^2 > \varphi(mv_0)$ with φ as in theorem 5.6.2 then $M_{H,A}^s(mv)$ is nonempty.*

2. *Let $m = 1$ and assume $v^2 > \varphi(v_0)$. Then $M_{H,A}(v)$ is an irreducible symplectic manifold and deformation equivalent to $\mathrm{Hilb}^{\frac{v^2}{2}+1}(X)$.*

3. *Let $m = 1$ and $v = (2, v_1, \frac{v_1^2-2}{4})$. Then $M_{H,A}(v)$ is birational to $\mathrm{Hilb}^2(X)$ or to X^2 or it is empty. In the first case $M_{H,A}(v)$ is an irreducible symplectic manifold and deformation equivalent to $\mathrm{Hilb}^2(X)$, in the second case it cannot be an irreducible symplectic manifold.*

4. *Let $m = 2$, $v^2 = 2$ and $v_0 = 1$. Then the symplectic varieties $M_{H,A}(2v)$ and $M_B(2v)$ are birational for a suitable $2v$-general ample divisor B on X, hence also any symplectic resolutions M of $M_{H,A}(2v)$ and M' of $M_B(2v)$.*

   If furthermore M or M' is an irreducible symplectic manifold then both are irreducible symplectic and deformation equivalent.

5. *Let $M_{H,A}^s(mv)$ be nonempty, $m \geq 2$ and $(mv)^2 \geq 16$, and assume $v_0 = 1$ or $v^2 > \varphi(v_0)$. Then $M_{H,A}(mv)$ is a singular locally factorial (and therefore $\mathbb{Q}$-factorial) projective symplectic variety with only terminal singularities, and in particular, there is no projective symplectic resolution.*

5.7. MORE RESULTS ON $M_{H,A}(V)$

Proof.

1. This is theorem 5.6.2 together with lemma 5.1.2.

2. This is corollary 5.4.2 using theorem 5.6.2 and lemma 5.1.2.

3. As v is primitive one has $M^s_{H,A}(v) = M_{H,A}(v)$ by proposition 5.1.4. Let $M_{H,A}(v)$ be nonempty. If $M^s_{H,A}(v)$ contains a sheaf F that is stable with respect to a v-general ample divisor B then corollary 5.4.2 yields the claim. If such a sheaf does not exist and B is any v-general ample divisor then $M^s_{H,A}(v)$ is birational to $M_B(v') \times M_B(v'')$ with $v' = (1, v'_1, v'_2)$, $v'' = (1, v''_1, v''_2)$ and $v'^2 + v''^2 = 0$ by corollary 5.6.5 as $M_{H,A}(v)$ is irreducible by corollary 5.2.4. In particular, $-2 \leq v'^2 = -v''^2 \leq 2$. Hence $M_B(v') \times M_B(v'')$ is either isomorphic to $\mathrm{Hilb}^2(X)$ or to $X \times X$. The additional statement follows by theorem 1.1.4.

4. The Mukai vector $w := 2v$ satisfies $8 = w^2 > \varphi(w_0)$, hence by theorem 5.6.2 and lemma 5.1.2 there is an (H, A)-stable and B-stable sheaf with Mukai vector w, and by corollary 5.4.2 $M_{H,A}(w)$ and $M_B(w)$ are birational. The statement on deformation equivalence follows by theorem 1.1.4 as usual.

5. If $v_0 \geq 2$ then by theorem 5.6.2 together with lemma 5.1.2 there is an (H, A)-stable sheaf with Mukai vector v. If $v_0 = 1$ then there is an A-stable sheaf with Mukai vector v by proposition 1.6.4 which is (H, A)-stable by corollary 4.2.17. In particular, $M^s_{H,A}(v) \neq \emptyset$, so together with lemma 3.2.1 the assumption of theorem 5.2.5 item 3 holds. □

The following question remains open:

Question 5.7.2 *Let X be a projective K3 surface, $v = (v_0, v_1, v_2) \in \Lambda(X)$ with $v_0 \geq 2$ and $2 \leq v^2 \leq \varphi(v_0)$ with φ as in theorem 5.6.2, H a not v-general ample divisor on X and A a v-general ample divisor on X. Assume that v is primitive or that $v^2 = 8$.*

1. *Is there an (H, A)-stable sheaf with Mukai vector v?*

2. Is there such a sheaf that is additionally B-stable for some v-general ample divisor B?

A positive answer for both items would exclude new examples of projective irreducible symplectic manifolds as we have seen.

5.8 Results on $\overline{M_H^s(v)}$

We return to the moduli space of H-semistable torsion free sheaves on a K3 surface and include the more explicit results.

Theorem 5.8.1 *Let X be a projective K3 surface, $v = (v_0, v_1, v_2) \in \Lambda(X)$ primitive with $v_0 > 0$, $m \in \mathbb{N}$ and H an ample divisor on X. Furthermore, assume that $M_H^s(mv)$ is nonempty. Then one has $v^2 \geq -2$, and in the case of equality one has $m = 1$ and $M_H(v)$ consists of a reduced point. Let now $v^2 \geq 0$.*

1. *Let $m = 1$ or $(mv)^2 \leq 14$. Then there is a projective symplectic resolution $M \to \overline{M_H^s(mv)}$. If H is not mv-general then M can be chosen to be a symplectic resolution of $M_{H,A}(mv)$, where A is an mv-general ample divisor.*

   Moreover, if M can be chosen to be irreducible symplectic then it is unique up to deformation.

2. *Let $m \geq 2$ and $(mv)^2 \geq 16$. If H is mv-general or $v_0 = 1$ or $v^2 > \varphi(v_0)$ with φ as in theorem 5.6.2 then there is a singular locally factorial (and therefore $\mathbb{Q}$-factorial) projective symplectic terminalisation of $\overline{M_H^s(mv)}$, and in particular, there is no projective symplectic resolution of $\overline{M_H^s(mv)}$.*

Proof. If H is mv-general then this holds by proposition 1.6.4 using lemma 3.2.1 for the case differentiation. Assume that H is not mv-general. The first part is proposition 1.3.7, and for $v^2 \geq 0$ the other statements are given by theorems 5.2.5 and 5.7.1 together with proposition 5.3.2. □

Let us conclude the discussion with a look at small ranks.

5.8. RESULTS ON $\overline{M_H^S(V)}$

Proposition 5.8.2 *Let X be a projective K3 surface, $v = (v_0, v_1, v_2) \in \Lambda(X)$ with $1 \leq v_0 \leq 3$, H an ample divisor on X, $M \subset M_H(v)$ an irreducible component, $\tilde{M}$ a projective irreducible symplectic manifold and $\tilde{M} \to M$ a projective birational morphism. Then $\tilde{M}$ is deformation equivalent to some resolution of some $M_B(w)$ with $w \in \Lambda(X)$ and B some w-general ample divisor or one has $v_0 = 3$, M contains stable sheaves and $\tilde{M}$ is deformation equivalent to some resolution of $M_{H,A}(v)$ for some v-general A, $2 \leq v^2 \leq \varphi(v_0)$ and there is no (H,A)-stable sheaf with Mukai vector v that is μ_B-semistable for any v-general B.*

Proof. Use proposition 5.8.1 together with theorem 5.7.1, corollary 5.4.2, lemma 5.1.3 and corollary 5.3.3. □

In particular, no new examples of a projective irreducible symplectic manifold arise from moduli spaces of semistable rank two sheaves.

Chapter 6

The relation to twisted stability

6.1 Twisted stability

Let X be a nonsingular projective variety of dimension d over an algebraically closed field k, F a coherent sheaf on X, H an ample line bundle on X and $D \in \mathrm{NS}(X)_{\mathbb{Q}} := \mathrm{NS}(X) \otimes \mathbb{Q}$. In our notation we might occasionally omit the map $c_1 : \mathrm{Pic}(X) \to \mathrm{NS}(X)$ when applied to a line bundle.

Definition 6.1.1 *1. The D-twisted Euler characteristic of F is*

$$\chi^D(F) := \int_X \mathrm{ch}(F).\exp(D).\mathrm{td}(X),$$

2. the D-twisted Hilbert polynomial of F is

$$P_H^D(F)(n) := \chi^D(F(nH)) = \chi^{D+nH}(F)$$

3. and the reduced D-twisted Hilbert polynomial of F is

$$p_H^D(F) := \frac{P_H^D(F)}{\alpha_{\dim F}^H(F)},$$

where $\alpha_{\dim F}^H(F)$ is the multiplicity of F, see section 4.2.

4. F is D-twisted H-(semi)stable if F is pure and for all nontrivial saturated proper subsheaves $E \subset F$ one has $p_H^D(E) (\leq) p_H^D(F)$.

For $D = 0$ the twisted notions coincide with the usual notions of Euler characteristic, (reduced) Hilbert polynomial and (semi)stability, and for a line bundle L one has $\chi^L(F) = \chi(F \otimes L)$ by Riemann-Roch. Hence using lemma 1.4.11 one has the following lemma.

Lemma 6.1.2 *Let L be a line bundle. Then F is L-twisted H-(semi)stable if and only if $F \otimes L$ is H-(semi)stable.*

In order to get more explicit formulae let X be a nonsingular projective surface over k, K_X its canonical divisor, and E and F two-dimensional sheaves. By Riemann-Roch one has

$$\chi^D(E) = \frac{\text{rk } E}{2} D^2 + \left(c_1(E) - \frac{\text{rk } E}{2} K_X \right).D + \chi(E)$$

and therefore

$$\frac{\chi^D(E)}{\text{rk } E} = \frac{1}{2} D^2 + \mu_D(E) - \frac{1}{2} K_X.D + \frac{\chi(E)}{\text{rk } E}$$

and

$$\frac{\chi^D(F)}{\text{rk } F} - \frac{\chi^D(E)}{\text{rk } E} = \mu_D(F) - \mu_D(E) + \frac{\chi(F)}{\text{rk } F} - \frac{\chi(E)}{\text{rk } E}.$$

The reduced D-twisted Hilbert polynomial is

$$p_H^D(E) = \frac{\chi^{D+nH}(E)}{H^2 \text{rk } E},$$

so one has

$$H^2 \left(p_H^D(F) - p_H^D(E) \right)$$
$$= (\mu_H(F) - \mu_H(E)) n + \mu_D(F) - \mu_D(E) + \frac{\chi(F)}{\text{rk } F} - \frac{\chi(E)}{\text{rk } E}. \quad (6.1)$$

6.2 Two-dimensional sheaves on a K3 surface

We restrict further to X being a projective K3 surface. Let $v = (v_0, v_1, v_2) \in \Lambda(X)$ with $v_0 \geq 2$, H an ample divisor lying on exactly one v-wall W and A a v-general ample divisor lying in one of the chambers touching H.

Definition 6.2.1 *For a nontrivial saturated subsheaf $E \subset F$ of a μ_H-semistable sheaf F with $v(F) = v$, $\mu_H(E) = \mu_H(F)$, and*

$$\frac{c_1(E)}{\operatorname{rk} E} \neq \frac{c_1(F)}{\operatorname{rk} F},$$

we define the hyperplane

$$\left\{ z \in \operatorname{NS}(X)_{\mathbb{Q}} \mid \frac{\chi^z(E)}{\operatorname{rk} E} = \frac{\chi^z(F)}{\operatorname{rk} F} \right\},$$

and call it a v-miniwall. The connected components of the complement of all v-miniwalls are called v-minichambers. Both notions are inspired by Ellingsrud and Göttsche.

In the following we omit the v-prefix as it is fixed for the whole section.

Proposition 6.2.2 *The number of miniwalls is finite and the miniwalls are parallel to W. Let $D, D' \in \operatorname{NS}(X)_{\mathbb{Q}}$. Then the set of D-twisted H-semistable sheaves is the same as the set of D'-twisted H-semistable sheaves if and only if D and D' belong to the same v-minichamber or v-miniwall.*

Proof. [MW97] proposition 3.5. □

Lemma 6.2.3 *Let D be in a minichamber and F a D-twisted H-semistable sheaf with $v(F) = v$. Then for every nontrivial saturated subsheaf $E \subset F$ with $p_H^D(E) = p_H^D(F)$ one has*

$$\frac{v(E)}{\operatorname{rk} E} = \frac{v(F)}{\operatorname{rk} F}.$$

Proof. Let $E \subset F$ be a nontrivial saturated subsheaf with $p_H^D(E) = p_H^D(F)$. In particular, one has $\mu_H(E) = \mu_H(F)$ and

$$\frac{\chi^D(E)}{\operatorname{rk} E} = \frac{\chi^D(F)}{\operatorname{rk} F}.$$

As D is not contained in a miniwall, one has

$$\frac{c_1(E)}{\operatorname{rk} E} = \frac{c_1(F)}{\operatorname{rk} F}$$

and thus also

$$\frac{\chi(E)}{\operatorname{rk} E} = \frac{\chi(F)}{\operatorname{rk} F}.$$

□

Lemma 6.2.4 *Let L be in a minichamber C, $L' \in \bar{C}$, and F a coherent sheaf on X with $v(F) = v$.*

1. *If F is L-twisted H-semistable then it is also L'-twisted H-semistable.*

2. *If F is L'-twisted H-stable then it is also L-twisted H-stable.*

Proof. Let $E \subset F$ be a nontrivial saturated proper subsheaf. If $\mu_H(E) < \mu_H(F)$ then clearly $p_H^D(E) < p_H^D(F)$ for any $D \in \operatorname{NS}(X)_\mathbb{Q}$. So let $\mu_H(E) = \mu_H(F)$. We define the map

$$f : \bar{C} \to \mathbb{Q}, D \mapsto \left(\frac{c_1(E)}{\operatorname{rk} E} - \frac{c_1(F)}{\operatorname{rk} F} \right).D + \frac{\chi(E)}{\operatorname{rk} E} - \frac{\chi(F)}{\operatorname{rk} F}.$$

If $\frac{c_1(E)}{\operatorname{rk} E} = \frac{c_1(F)}{\operatorname{rk} F}$ then f is independent of D. So let $\frac{c_1(E)}{\operatorname{rk} E} \neq \frac{c_1(F)}{\operatorname{rk} F}$. Then $f \neq 0$ on the whole minichamber C. We distinguish the two cases from above.

1. Let F be L-twisted H-semistable. Then $f < 0$ on C, hence $f \leq 0$.

2. Let F be L'-twisted H-stable. Then $f(L') < 0$, hence $f < 0$ on an open subset containing L', which in turn yields $f < 0$ on C.

□

6.2. TWO-DIMENSIONAL SHEAVES ON A K3 SURFACE

Proposition 6.2.5 *Let L be in a minichamber C, $L' \in \partial C$, and F a coherent sheaf on X with $v(F) = v$. The vector space generated by the wall W divides $\mathrm{NS}(X)_{\mathbb{Q}}$ into two open half spaces, one of them containing $L - L'$. Choose A in the neighbouring chamber of W contained in the other half space. Then F is L-twisted H-(semi)stable if and only if it is L'-twisted H-semistable and for all nontrivial saturated proper subsheaves $E \subset F$ with $p_H^{L'}(E) = p_H^{L'}(F)$ one has $\mu_A(E) \,(\geq)\, \mu_A(F)$.*

Proof. Let $E \subset F$ be a nontrivial saturated proper subsheaf. If one has $\mu_H(E) < \mu_H(F)$ then clearly $p_H^D(E) < p_H^D(F)$ for any $D \in \mathrm{NS}(X)_{\mathbb{Q}}$. So let $\mu_H(E) = \mu_H(F)$. Thus

$$p_H^L(E) - p_H^L(F) - (p_H^{L'}(E) - p_H^{L'}(F)) = \left(\frac{c_1(E)}{\mathrm{rk}\, E} - \frac{c_1(F)}{\mathrm{rk}\, F}\right).(L - L')\frac{1}{H^2} \quad (6.2)$$

by equation (6.1). If $\frac{c_1(E)}{\mathrm{rk}\, E} = \frac{c_1(F)}{\mathrm{rk}\, F}$ then

$$p_H^L(E) - p_H^L(F) = p_H^{L'}(E) - p_H^{L'}(F)$$

and $\mu_A(E) = \mu_A(F)$, so we assume

$$\frac{c_1(E)}{\mathrm{rk}\, E} - \frac{c_1(F)}{\mathrm{rk}\, F} \neq 0,$$

which thus defines the wall W. In particular, the sign of

$$\left(\frac{c_1(E)}{\mathrm{rk}\, E} - \frac{c_1(F)}{\mathrm{rk}\, F}\right).(L - L') \neq 0$$

is opposite to the sign of $\mu_A(E) - \mu_A(F)$.

1. Assume that F is L-twisted H-semistable and thus also L'-twisted H-semistable by lemma 6.2.4. If furthermore $p_H^{L'}(E) = p_H^{L'}(F)$ then equation

(6.2) yields

$$p_H^L(E) - p_H^L(F) = \left(\frac{c_1(E)}{\text{rk } E} - \frac{c_1(F)}{\text{rk } F}\right).(L - L')\frac{2}{H^2},$$

which is negative, hence $\mu_A(E) > \mu_A(F)$.

2. Assume that F is L'-twisted H-semistable, i.e. $p_H^{L'}(E) \leq p_H^{L'}(F)$. If one has strict inequality then by the same argument as in lemma 6.2.4 one has $p_H^L(E) < p_H^L(F)$. So let's assume equality. Then $\mu_A(E) \geq \mu_A(F)$ and thus

$$p_H^L(E) - p_H^L(F) = \left(\frac{c_1(E)}{\text{rk } E} - \frac{c_1(F)}{\text{rk } F}\right).(L - L')\frac{2}{H^2} < 0.$$

□

The following statement, at least the part on semistability, is already known to Matsuki and Wentworth, as it can be found in [MW97] theorem 4.1, part i.

Corollary 6.2.6 *Let A be an ample divisor in a chamber touching H and $L \in \text{Pic}(X)$ lying on a miniwall. The vector space generated by the wall W divides $\text{NS}(X)_{\mathbb{Q}}$ into two open half spaces, one of them containing A. Choose D in one of the minichambers touching L such that $D - L$ is in the other half space. Then a coherent sheaf F with $v(F) = v$ is D-twisted H-(semi)stable if and only if $F \otimes L$ is (H, A)-(semi)stable.*

Proof. This follows from proposition 6.2.5, lemma 6.1.2, the characterisation in proposition 4.2.15 and the explicit inequalitites in lemma 5.1.1. □

If in the situation of the corollary D can be chosen to be the first Chern class of a line bundle L' then one has an isomorphism

$$M_{H,A}(v.\exp(L)) \cong M_H(v.\exp(L')).$$

In general $M_{H,A}(v.\exp(L))$ can be seen as a moduli space for D-twisted H-semistable sheaves with Mukai vector v. For more details on this see [MW97].

Bibliography

[Bea83]   Arnaud Beauville, *Variétés Kähleriennes dont la première classe de Chern est nulle*, J. Differential Geom. **18** (1983), no. 4, 755–782 (1984).

[Bea00]   _____, *Symplectic singularities*, Invent. Math. **139** (2000), no. 3, 541–549.

[Bea10]   _____, *Holomorphic symplectic geometry: a problem list*, arXiv:1002.4321v1 [math.AG] (2010).

[Ber55]   Marcel Berger, *Sur les groupes d'holonomie homogène des variétés à connexion affine et des variétés riemanniennes*, Bull. Soc. Math. France **83** (1955), 279–330.

[BHPVdV04] Wolf P. Barth, Klaus Hulek, Chris A. M. Peters, and Antonius Van de Ven, *Compact complex surfaces*, second ed., Ergebnisse der Mathematik und ihrer Grenzgebiete. 3. Folge. A Series of Modern Surveys in Mathematics [Results in Mathematics and Related Areas. 3rd Series. A Series of Modern Surveys in Mathematics], vol. 4, Springer-Verlag, Berlin, 2004.

[Bog74]   Fedor Alekseevich Bogomolov, *On the decomposition of Kähler manifolds with trivial canonical class*, Mat. Sb. **93(135)** (1974), no. 4, 573–575.

BIBLIOGRAPHY

[BS10]   Samuel Boissière and Olivier Serman, *Sur le produit de variétés localement factorielles ou Q-factorielles*, preprint (2010).

[Fog68] John Fogarty, *Algebraic families on an algebraic surface*, Amer. J. Math **90** (1968), 511–521.

[Gro61] A. Grothendieck, *Éléments de géométrie algébrique. III. Étude cohomologique des faisceaux cohérents. I*, Inst. Hautes Études Sci. Publ. Math. (1961), no. 11, 167.

[Gro65] _____, *Éléments de géométrie algébrique. IV. Étude locale des schémas et des morphismes de schémas. II*, Inst. Hautes Études Sci. Publ. Math. (1965), no. 24, 231.

[Har77] Robin Hartshorne, *Algebraic geometry*, Springer-Verlag, New York, 1977, Graduate Texts in Mathematics, No. 52.

[HL97]   Daniel Huybrechts and Manfred Lehn, *The geometry of moduli spaces of sheaves*, Aspects of Mathematics, E31, Friedr. Vieweg & Sohn, Braunschweig, 1997.

[Huy99] Daniel Huybrechts, *Compact hyper-Kähler manifolds: basic results*, Invent. Math. **135** (1999), no. 1, 63–113.

[KLS06] D. Kaledin, M. Lehn, and Ch. Sorger, *Singular symplectic moduli spaces*, Invent. Math. **164** (2006), no. 3, 591–614.

[KM98]   János Kollár and Shigefumi Mori, *Birational geometry of algebraic varieties*, Cambridge Tracts in Mathematics, vol. 134, Cambridge University Press, Cambridge, 1998, With the collaboration of C. H. Clemens and A. Corti, Translated from the 1998 Japanese original.

[Leh02] Manfred Lehn, *Symplectic moduli spaces*, Trieste Lecture Notes, 2002.

BIBLIOGRAPHY

[LS06]   Manfred Lehn and Christoph Sorger, *La singularité de O'Grady*, J. Algebraic Geom. **15** (2006), no. 4, 753–770.

[Muk84] Shigeru Mukai, *Symplectic structure of the moduli space of sheaves on an abelian or $K3$ surface*, Invent. Math. **77** (1984), no. 1, 101–116.

[Muk87] S. Mukai, *On the moduli space of bundles on $K3$ surfaces. I*, Vector bundles on algebraic varieties (Bombay, 1984), Tata Inst. Fund. Res. Stud. Math., vol. 11, Tata Inst. Fund. Res., Bombay, 1987, pp. 341–413.

[Mum99] David Mumford, *The red book of varieties and schemes*, expanded ed., Lecture Notes in Mathematics, vol. 1358, Springer-Verlag, Berlin, 1999, Includes the Michigan lectures (1974) on curves and their Jacobians, With contributions by Enrico Arbarello.

[MW97]   Kenji Matsuki and Richard Wentworth, *Mumford-Thaddeus principle on the moduli space of vector bundles on an algebraic surface*, Internat. J. Math. **8** (1997), no. 1, 97–148.

[Nak99] Hiraku Nakajima, *Lectures on Hilbert schemes of points on surfaces*, University Lecture Series, vol. 18, American Mathematical Society, Providence, RI, 1999.

[Nam01] Yoshinori Namikawa, *A note on symplectic singularities*, arXiv:math/0101028v1 [math.AG] (2001).

[Nam06] _____, *On deformations of $\mathbb{Q}$-factorial symplectic varieties*, J. Reine Angew. Math. **599** (2006), 97–110.

[O'G96] Kieran G. O'Grady, *Moduli of vector bundles on projective surfaces: some basic results*, Invent. Math. **123** (1996), no. 1, 141–207.

[O'G99]   ———, *Desingularized moduli spaces of sheaves on a K3*, J. Reine Angew. Math. **512** (1999), 49–117.

[O'G03]   ———, *A new six-dimensional irreducible symplectic variety*, J. Algebraic Geom. **12** (2003), no. 3, 435–505.

[Yau78]   Shing Tung Yau, *On the Ricci curvature of a compact Kähler manifold and the complex Monge-Ampère equation. I*, Comm. Pure Appl. Math. **31** (1978), no. 3, 339–411.

[Yos00]   Kōta Yoshioka, *Irreducibility of moduli spaces of vector bundles on k3 surfaces*, arXiv:math/9907001v2 [math.AG] (2000).

[Yos01]   ———, *Moduli spaces of stable sheaves on abelian surfaces*, Math. Ann. **321** (2001), no. 4, 817–884.

Acknowledgement

The author would like to express his gratitude to his doctoral advisor for the optimal support, his suggestions and his encouragement. He also thanks all colleagues for stimulating discussions and his family and friends. Most part of this work was supported by the SFB/TR 45 of the DFG.

I want morebooks!

Buy your books fast and straightforward online - at one of world's fastest growing online book stores! Environmentally sound due to Print-on-Demand technologies.

Buy your books online at
www.morebooks.shop

Kaufen Sie Ihre Bücher schnell und unkompliziert online – auf einer der am schnellsten wachsenden Buchhandelsplattformen weltweit! Dank Print-On-Demand umwelt- und ressourcenschonend produziert.

Bücher schneller online kaufen
www.morebooks.shop

KS OmniScriptum Publishing
Brivibas gatve 197
LV-1039 Riga, Latvia
Telefax: +371 686 204 55

info@omniscriptum.com
www.omniscriptum.com

Printed by Books on Demand GmbH, Norderstedt / Germany